EVOLUTIONARIES

EVOLUTIONARIES

Unlocking the Spiritual and Cultural Potential of Science's Greatest Idea

CARTER PHIPPS

HARPER ● PERENNIAL

NEW YORK ● LONDON ● TORONTO ● SYDNEY ● NEW DELHI ● AUCKLAND

HARPER ⬤ PERENNIAL

HarperCollins books may be purchased for educational, business, or sales promotional use. For information please write: Special Markets Department, HarperCollins Publishers, 10 East 53rd Street, New York, NY 10022.

FIRST EDITION

Designed by Michael P. Correy

Library of Congress Cataloging-in-Publication Data is available upon request.

ISBN 978-0-06-191613-7

12 13 14 15 16   ID/RRD   10 9 8 7 6 5 4 3 2

To the great pioneers in evolutionary
science, philosophy, and spirituality
whose vision, dedication, perseverance, and
faith created new pathways for us all.

We are moving!

—*Pierre Teilhard de Chardin*, The Future of Man

CONTENTS

Introduction   xi

Part I: Reexamining Evolution

Prologue: An Evolutionary Vision   3

1   Evolution: A New Worldview   7

2   Breaking the Spell of Solidity   21

3   What Is an Evolutionary?   31

Part II: Reinterpreting Science

4   Cooperation: A Sociable Cosmos   49

5   Directionality: The Road to Somewhere   71

6   Novelty: The God Problem   101

7   Transhumanism: An Exponential Runway   125

Part III: Recontextualizing Culture

8   The Internal Universe   155

CONTENTS

9   Evolving Consciousness: The Inside Story   181

10 Spiral Dynamics: The Invisible Scaffolding of Culture   211

11 An Integral Vision   233

Part IV: Reenvisioning Spirit

12 Evolutionary Spirituality: A New Orientation   263

13 Conscious Evolution: Our Moment of Choice   285

14 The Evolution of Enlightenment   317

15 An Evolving God   339

16 Pilgrims of the Future   365

Acknowledgments   373

Notes   375

Index   391

Do you believe the Bible, or do you believe in evolution?" The question came in an urgent whisper, passed around my eighth-grade science class with the guilty subterfuge of a dirty joke. I was thirteen years old and already I understood that there were two kinds of people in my small town on the edge of the Bible Belt: those who believed and those who didn't.

I'm not talking about God—everyone believed in God. I'm talking about evolution.

Little did I realize, at that tender age, that I was confronting one of the most profound and enduring existential dilemmas of my time and culture. A few decades later, in the brave new world of the twenty-first century, that whispered question, far from being an out-dated relic, is emblazoned across the covers of the most mainstream magazines: "God vs. Darwin," "Science vs. Spirit," "The Plot to Kill Evolution . . ." Indeed, evolution, these days, instead of just being a sci-entific term denoting a biological explanation for the origin of life, has become almost a pseudonym for the endemic culture wars that simmer under the surface of American society and occasionally flare up into full-fledged showdowns that temporarily consume the consciousness of our mass media. The debate over evolution, we are told, is a war between irrevocably opposed camps—between those who look at the natural world and see the handiwork of a divine intelligence and those who look at the world and see only impersonal processes and meaningless manifestations of matter. In fact, judging from the headlines alone, one would have to conclude that we live in a world where God and evolution are mutually exclusive.

But like my Oklahoma hometown, America is not that simple. If the pollsters are to be trusted, 91 percent of Americans believe in God. And 50 percent accept the scientific theory of evolution. You just have to do the math to see that the black-and-white image of a polarized nation that makes books and magazines fly off the shelves is concealing at least a few shades of gray. In fact, as I've found out in my extensive research over the last decade, it conceals a whole spectrum of colors.

Personally, I always knew exactly where I stood in the evolution wars. I grew up a lover of knowledge, of science, and Carl Sagan was my childhood hero. Although I was only twelve years old when *Cosmos*, Sagan's signature PBS show on science and space, aired, I watched each episode as if it were a religious revelation. After the show, in the quiet of my bedroom, I would read the accompanying book cover to cover, letting my mind roam across the universe, imagining new worlds and new forms of matter and life. I was awed by the power of black holes, humbled by the size of quasars, and inspired by the thought that one day humans might travel among the stars.

As the years passed, I always retained that basic passion and appreciation for the ever-developing knowledge of science. There were certainly those in my hometown who saw a godless, evil half-truth in the declarations of Darwin and his intellectual descendants, but for me, my friends, and my family, evolution—like the rest of science—was just another fascinating and noncontroversial fact of life.

My relationship with religion, however, has always been more complex. Raised in an intellectually adventurous family, I never thought of myself as a religious person. Yes, I went to church more Sundays than not—Presbyterian was our family denomination. But church was simply a respected and honored part of the multilayered social fabric of life in a small town, not something to be taken too seriously, and certainly not something to be fanatical about. Such things were better left to the Baptists, or the overly religious folk down at the Church of Christ. By my late teenage years, however, my outlook had begun to shift. Fate seemed to instill in my young heart a passion for discovering a deeper meaning to life, and a yearning for spiritual fulfillment began to well up in my consciousness. By the end of my time at university, spirituality—and, more specifically, Eastern philosophy and meditation—had become a focal point of my life. Finally, at the age of twenty-two, in 1990, only two weeks after I graduated early from Oklahoma University, I left the world I knew behind—my honors degree, my fraternity brothers, plans for law school—and boarded a plane for the East to dedicate myself to the pursuit of wisdom, truth, and spiritual knowledge.

A decade later, I found myself in a unique position—editor at one of the most influential and progressive spiritual and philosophical magazines in America, *EnlightenNext* (formerly *What Is Enlightenment?*) And my early passion for science began to play a key role in my new career. In fact, the relationship between science and spirit became one of the most important areas of inquiry for a magazine dedicated to defining a post-traditional spiritual and philosophical worldview suited to a rational age. In the course of my work for *EnlightenNext*, I was privileged to be able to meet and interview some of the most impressive spiritual leaders and brilliant scientific pioneers alive today. I had the chance to examine where they stood on many of the most important questions confronting human culture. And I have learned that evolution is not only a line in the sand between science and faith. It is also, I have been surprised to find, a bridge that connects them.

The journey that led to the creation of this book was what author Steven Johnson calls "a slow hunch"—a gradual epiphany that coalesced around a number of important insights and encounters. As I interviewed many individuals who were breaking new ground in science, spirituality, philosophy, psychology, and even traditional religion, I noticed that a common theme was emerging—evolution. They were informed and inspired by science's revelation of our biological and cosmic history, and in some cases explicitly championing the notion of evolution as a new way of making meaning out of life in the twenty-first century. In the hands of some, this evolutionary inspiration leaned toward secularism and humanism, and in the hands of others it leaned toward pantheism or even theism. But they all shared a common evolutionary context for interpreting their fields. We were witnessing, I eventually concluded, the birth of an authentic new spiritual and philosophical worldview. This new worldview was science-friendly and placed questions of human purpose and meaning fully in the context of an evolutionary cosmos. It has been slowly emerging in the culture over the last two centuries but has picked up steam in the last two decades.

To be sure, there is still little cultural mind-space in mainstream America for the emergence of evolutionarily inspired forms of meaning and purpose. Let's be honest: God vs. Darwin has great media resonance. But more and more intelligent people are beginning to question this easy dichotomy. In fact, long before I fully understood what a critical role evolution would play in developing a worldview that is adequate to meet the demands of a new century, I knew that the choice the popular media so often presents—believe in God or believe in evolution; embrace atheism or be lumped in with intelligent design—is a false choice. It is not actually a choice between the spiritual impulse and the scientific impulse but between two worldviews: one that believes in the ultimate primacy of matter and one that believes in the ultimate primacy of an ancient god. And I, for one, don't believe in either.

So, together with my fellow editors at *EnlightenNext*, I set out on the research journey that eventually led to this book. My initial intention, which resulted in a feature article, was to uncover what I called "the *real* evolution debate"—to chart those scientific, philosophical, and metaphysical perspectives that are causing us to redefine the nature of the evolutionary process and to rethink our conclusions about where we come from, who we are, and where we might be going. As I delved behind the polarizing headlines, I uncovered an extraordinary world where conventional ideas about both science and religion are being turned upside down, and scientists, philosophers, and theologians are negotiating, in ever-surprising ways, what Plutarch described as the "difficult course between the precipice of godlessness and the marsh of superstition."

Perhaps no single entity has done more to feature the thoughts and ideas that make up the basic framework of a new evolution-inspired worldview over the last decades than *EnlightenNext* magazine. Though small in scale, the magazine, during its almost twenty-year history, played a catalytic role in featuring the people and perspectives associated with an evolutionary understanding of the world. As executive editor, I had the unique opportunity to play several roles in this movement—journalist, critic, witness, and also creative participant. My position has afforded me the opportunity to engage with many of the remarkable scientists, religious thinkers, philosophers, spiritual visionaries, psychologists, researchers, and theorists who are spearheading this novel perspective. It is to those many thousands of hours of enlightening conversation, dialogue, and discussion that the contents of this book owe their debt. I am a firm believer in the idea that innovation and creativity is as much a function of the right kind of relationships as it is of a particular kind of individual vision. In that respect, I have been blessed to be part of an extremely stimulating network of friends, colleagues, and collaborators.

Evolution is one of those bellwether ideas that is able to uniquely track the currents of our cultural zeitgeist, simply because its roots reach so deep into the way we understand reality. It is not hyperbole to say that how we think about evolution profoundly affects how we think about life, the universe, and everything. That is why it is a critical pillar in the work to form a new worldview that can meet the demands of the twenty-first century. I'm not alone in this conviction. This book brings together a diverse but interconnected ecosystem of theorists, researchers, teachers, and philosophers who are, each in their own way, helping to contribute to this critical cultural project. As the contours of this evolutionary worldview become clear, I am confident that it will help us to find new and creative responses to the many challenges of life in this complex and rapidly changing world that we have inherited. How we think about evolution is foundational to the kinds of visions that we hold for our collective future. It shapes our understanding of who we are today, how we got here, and what our role is in creating the world of tomorrow. Confronted by the unprecedented challenges of a globalizing, environmentally threatened, culturally dissonant world, nothing could be more critical. Paradoxically, the debate about our origins is also a cultural referendum on our future.

PART I
REEXAMINING EVOLUTION

An Evolutionary Vision

On November 24, 1859, a little-known biologist from England quietly published a book introducing a significant new scientific theory, proposing that a process he termed "natural selection" could explain how human beings had evolved from other species. The title would soon become known the world over—*On the Origin of Species*. The first edition sold out within days, all 1,170 copies, and the rest, as they say, is history. . . .

One hundred years later, in 1959, this event had become reason for celebration. A number of leading evolutionary pioneers gathered together at the University of Chicago to commemorate the centennial of the publication of Charles Darwin's first book, spending several autumn days on the beautiful tree-lined campus paying homage to his unique genius and reflecting on the meaning of evolution. The star-studded interdisciplinary conference featured presentations from experts in the fields of biology, paleontology, anthropology, and even psychology. The best and brightest were in attendance, including legendary evolutionary biologist Ernst Mayr and geneticist Theodore Dobzhansky, who each shared their wisdom with the assembled audience. Even Darwin's grandson was present.

But perhaps the most famous guest of all was the grandson of another great evolutionist, the English biologist Thomas Henry

Huxley, one of the early supporters of Darwin's revolutionary theory. Julian Huxley, his descendant, was a brilliant scientist, humanist, and world-renowned intellectual. As he ascended the podium to address the international audience, expectations ran high. Here was a man who had worked to convince the world that Darwin's natural selection was a driving force of evolutionary change. The audience would have also known Huxley for his humanitarian ideals, which had helped inspire the great humanist movement, the twentieth century's intellectual alternative to religious faith. Some may have been aware of Huxley's interest in the existential implications of evolutionary theory, a passion that had led him to coin the phrase "We are evolution become conscious of itself." Perhaps some even knew him as the fiercely independent thinker who had endured the outrage of his secular-minded colleagues to write the introduction to the controversial book on religion and evolution, *The Phenomenon of Man*, by recently deceased Catholic priest and paleontologist Pierre Teilhard de Chardin. What would Huxley offer his audience on this momentous anniversary, when some of the greatest minds of the era had their attention trained on his pulpit?

Huxley's talk was called "The Evolutionary Vision," and he delivered it with an almost religious passion, attendees recalled. He suggested that religion as we knew it was dying, that "supernaturally centered" faiths were destined to decline, to deselect themselves out of existence like nonadaptive species in a hostile environment. "Evolutionary man can no longer take refuge from his loneliness in the arms of a divinized father figure whom he has himself created," Huxley claimed, "nor escape from the responsibility of making decisions by sheltering under the umbrella of Divine Authority, nor absolve himself from the hard task of meeting his present problems and planning his future by relying on the will of an omniscient, but unfortunately inscrutable, Providence." Huxley's words were strong, spoken with the conviction of one who had worked his whole life to free the human spirit from belief systems unsuited to the modern world. But before proclaiming the death of religion altogether, he added a notable line. "Finally," he concluded, "the evolutionary vision is enabling us to discern, however incompletely, the lineaments of the new religion that . . . will arise to serve the needs of the coming era."

For Huxley, evolution was not merely a final nail in the coffin of traditional religious belief. It represented much more than the victory of a scientific theory over the historical forces of superstition and ignorance. The triumph of evolution also pointed us toward the future— toward a post-traditional synthesis that would arise out of our new understanding of who we are and where we came from.

In the fall of 2009, I attended another conference at the University of Chicago, held exactly fifty years following the first gathering and one hundred and fifty years after the publication of *On the Origin of Species*. Like its predecessor, the event was also a meeting of some of evolutionary theory's brightest lights, and I was curious to see what the intellectual descendants of Huxley, Mayr, and Dobzhansky might have to say about the "evolutionary vision" fifty years on down the road.

I found the conference to be fascinating, the lectures and discussions on the latest findings in evolutionary science wonderfully informative. Religion, too, was a major subject of the day. Today's evolutionary scientists are veritably obsessed with their ongoing struggles against creationism and intelligent design; they are deeply vexed about the resistance to Darwin's ideas and biology's discoveries that still characterizes so many of today's religious communities. As someone who grew up in the Bible Belt, where such controversies rage unchecked, I understood and shared their concerns. But what of Julian Huxley's vision? What of his observation that a rich, novel kind of evolutionary knowledge might change our worldview, our sense of self and humanity's place in the scheme of things?

There was little to report from Chicago on that front. To hear the version of things presented in those hoary halls, there is the ongoing march of new science, the ongoing resistance of old-time religion, and that's about the extent of it. Admittedly, there was an occasional nod to the heroic attempt to reconcile evolution and faith, but no one was on the lookout for the emergence of a new evolution-inspired spirituality. No one was talking about the way in which evolutionary ideas might transform culture and human thought in the new century. In fact, it seemed that no one was paying much attention at all to the vision that Huxley had presented on that November day in 1959.

But just because they're not paying attention doesn't mean that there is nothing worth watching. Indeed, today Huxley's evolutionary vision is more culturally relevant than ever. It is living in the hearts and minds of thousands of individuals around the world who are experimenting with new cultural perspectives, new philosophical epiphanies, new spiritual ideals, new religious visions—all based around the idea of evolution. Sadly, these cultural pioneers were not invited to the 2009 conference in Chicago. To find them, we must travel outside the conventional walls of the academy and beyond the ancient structures of traditional religion. We must journey to the frothy frontiers of culture, to the border between convention and controversy where the next great cultural breakthroughs are struggling to be born. This is a book about the search for that evolutionary vision and a new kind of worldview based on it.

CHAPTER ONE

Evolution: A New Worldview

The most extraordinary fact about public awareness of evolution is not that 50 percent don't believe it but that nearly 100 percent haven't connected it to anything of importance in their lives. The reason we believe so firmly in the physical sciences is not because they are better documented than evolution but because they are so essential to our everyday lives. We can't build bridges, drive cars, or fly airplanes without them. In my opinion, evolutionary theory will prove just as essential to our welfare and we will wonder in retrospect how we lived in ignorance for so long.

—*David Sloan Wilson*, Evolution for Everyone

Evolution is a fact. Given the seemingly never-ending controversy surrounding biological science and all of its many discoveries regarding the origins of life, it's important to be clear right from the start. In this book, there is no controversy. I would say that I believe in evolution, only I don't think belief has anything to do with it. We don't say we believe the world is round—we know it is. Evolution is not a matter of faith; it is a matter of evidence, painstaking work, and breakthrough science. Any other conclusion stretches the bounds of credibility and retards the advance of knowledge. Evolution is simply true.

Now that I have stated my position clearly and unequivocally, let me confuse the matter. I also think that the discovery of evolution is the greatest cultural, philosophical, and spiritual event in the last few hundred years. I think its overall influence is destined, in the long run, to be seen alongside some of our culture's most significant inflection points—the birth of monotheism, the European Enlightenment, the industrial revolution.

Are you surprised that I used the word "spiritual"? Many, no doubt, will be. Hasn't the theory of evolution long been the number-one *enemy* of spirit in most religious circles? Isn't evolution the atheist's answer to religious faith, the "blind watchmaker" who has slowly fashioned life out of inanimate matter without any divine help? Didn't Darwin's paradigm-shattering revolution of natural selection and random mutation explain away God with one momentous insight into the workings of Mother Nature?

Yes, that is certainly the story as it is often told—the story that causes consternation in the classrooms of Kansas, inflames the passions of Christians from the rust belt to the Bible Belt, and riles up Muslims from Baghdad to Birmingham. But consider this: evolution was never merely a scientific idea. For that matter, it wasn't even Darwin's idea. Indeed, long before Darwin ever became fascinated by Galápagos finches, the notion of evolution was already at work in the culture of the nineteenth century, quietly subverting established categories of thought and changing religion, philosophy, and science, in unexpected and remarkable ways.

Please don't misunderstand me: I have the greatest respect for Darwin's seminal contribution. He was the critical match that turned the sparks of a subversive idea into a world-engulfing conflagration. When the historian Will Durant was asked toward the end of his life if Marx could be considered to be the person who had the most influence on the twentieth century, he begged to differ. Darwin, he replied, was even more influential than Marx. The theory of evolution was one of the primary drivers in the undermining of religious faith, and so Darwin's legacy loomed larger in the century's embrace of a more secular culture.

I concur with Durant's estimation of Darwin's transformative role. But I do think he overlooked a significant point. In the broadest sense, Darwin and Marx were both driven by the same fundamental idea—evolution! For better or worse, evolution was the context of each of their life's work. Marx was a student of one of the first great evolutionary philosophers, Hegel. And while Darwin, a meticulous collector of data, was focused on biology and Marx on political theory and economic history, both drank liberally from the same philosophical insight—that the given categories of life as it exists today are not static or fixed or unchanging, the "way things are," but rather are a momentary snapshot in an ongoing developmental process. They both saw through the illusion of permanence created by the seeming solidity of the objects of their respective work—for Darwin, the living world; for Marx, economic structures and historical processes—and understood that these are part of a deeper underlying process of evolution over time. I don't mean to endorse the ill-conceived nature of Marx's historical materialism and its tragic human consequences. My point is simply that the evolution revolution may ultimately prove bigger than even many of our most capable thinkers have yet grasped.

The idea of evolution, the basic notion of process, change, and development over time, is affecting much more than biology. It is affecting everything, from our perceptions of politics, economics, psychology, and ecology to our understanding of the most basic constituents of reality. It is helping to give birth to new philosophies and, I will argue, is the source of a new kind of spiritual revelation. The individuals featured in this book have all been inspired by Darwin and the enduring insights of the past century's breakthroughs in biology, genetics, and paleontology (some have even contributed to them). But they are also reaching beyond those laudable accomplishments to uncover new vistas. They are forging a rich, novel way of understanding the development of everything from the complex corridors of the human psyche to the outer reaches of the universe. They are drawn to discover the hidden structures deep within the interiors of the genome and also the hidden structures deep within the interiors of culture. Evolution, in these pages, is certainly about the birds and the bees, but it's also about culture, consciousness, and the cosmos. On this evolutionary journey, the insights of evolutionary science will always apply, but they will have to share the spotlight with pioneering thinkers and theories from a surprising diversity of fields.

I believe that our emerging understanding of evolution in all its many shapes and sizes and dimensions is so fundamental that it would be hard to overstate its significance. Taken as a whole, it will constitute the organizing principle of a new worldview, uniquely suited for the twenty-first century and beyond. The outlines of this worldview are still being formed—spurred on by new insights and breakthroughs in the development of science, psychology, sociology, technology, philosophy, and theology. This book is about that new worldview and the people who are consciously engaged in its creation. Working across vastly different contexts and disciplines, these individuals are united not by creed or belief system but by a broadly shared evolutionary vision and a care for our collective future. They are scientists and futurists, sociologists and psychologists, priests and politicians, philosophers and theologians. They share no common title. I call them Evolutionaries.

CONTEXT IS EVERYTHING

The great geneticist Theodore Dobzhansky once declared, "Nothing makes sense in biology except in light of evolution." In fact, before the discovery of evolution, biology was largely just a way of classifying species. Evolution was the unifying idea that placed it on the academic map as a coherent and legitimate science, and it became the context for so much of our understanding of the forms and features of life. Dobzhansky's insight, I am convinced, is applicable to a much broader sphere than biology. Arguably, it is true that *nothing in human culture makes sense except in light of evolution*. This book will make the case that our emerging understanding of evolution is so transformative that eventually every important area of human life will fall under its revelatory spell. It will change the way we think about life, culture, consciousness, even thinking itself—for the better. In fact, it already is.

Among the first to recognize the scale and significance of evolutionary thinking was the French Jesuit priest and paleontologist Pierre Teilhard de Chardin. "Is evolution a theory, a system, or a hypothesis?" he wrote in the early twentieth century. "It is much more: it is a general condition to which all theories, all hypotheses, all systems must bow and which they must satisfy henceforth if they are to be thinkable and true. Evolution is a light which illuminates all facts, a curve that all lines must follow."

If this quote seems to succumb to a kind of triumphalism, let me also try to balance the record by noting that we are still in the beginning phase of building a coherent worldview that incorporates the transformative insights of an evolutionary perspective. The people and ideas represented in this book do not capture anything in its final form. Like scholars in a young field of study, Evolutionaries have plenty of material to build on from those thinkers that have forged the way forward in the last two centuries. Yet things have not progressed far enough that ideas have become fixed, ways of thinking established, or agreements settled. We are somewhere between the initial visions of what's possible and established, accepted cultural truths. We are still in the Wild West phase of development, between the earliest pioneers exploring virgin land and homesteaders looking to settle down and build a new life on secured territory. As the many truths, insights, cultural perspectives, and attitudes that make up this new worldview eventually work themselves into the mainstream, they will have greater and greater effects on all aspects of human culture.

"A philosophy of this kind will not be made in a day," wrote the pioneering evolutionary philosopher and Nobel laureate Henri Bergson in his classic 1907 book *Creative Evolution*. "Unlike the philosophical systems properly so called, each of which was the individual work of a man of genius and sprang up as a whole, to be taken or left, it will only be built by the collective and progressive effort of many thinkers, of many observers also, competing, correcting, and improving one another."

EVERYTHING YOU THINK YOU KNOW ABOUT EVOLUTION MAY BE WRONG (OR INCOMPLETE)

When Darwin spoke about evolution, he meant *descent with modification*, the idea that all organisms are descended from a common ancestor. His theory of natural selection was a theory about the mechanisms of that modification. Over time, with the discovery of genetics as the agent of heredity, Darwin's theory became neo-Darwinism, or what is sometimes called the "modern synthesis"—the idea that evolution is driven by a combination of natural selection and random mutation. The basic proposition is that random mutation at the level of the gene produces novel forms and features in an organism. Those few features that are adaptive increase the "fitness" of an organism, allowing it to better survive and pass on its genes, thereby transferring those same features to the next generation. Those mutations that are nonadaptive or nonadvantageous, on the other hand, will naturally disappear because those organisms will be more likely to die out, making them less likely to pass on their genes to the next generation. The combination, therefore, of random mutation and differential selection became a powerful and inseparable scientific duo, playing a central role in the fledgling field of evolutionary biology. This new scientific consensus began to influence public perception, and eventually the term "evolution" and the scientific idea of natural selection and random mutation became almost synonymous.

But within the scientific community itself, such a close conflation is more myth than fact. Science is rarely so settled. Indeed, in recent decades, there have been new and interesting ideas making their way to the forefront of the field that raise important questions about our understanding of the history of life. Esteemed scientist and complexity theorist Stuart Kauffman, in his book *Reinventing the Sacred*, recently wrote, "There is considerable doubt . . . about the power and sufficiency of natural selection as the sole motor of evolution." His novel evolutionary ideas, along with those of the late biologist Lynn Margulis and others, have extended the science of evolution in significant ways, and I will further explore their work in Part I of this book. Moreover, the new science of epigenetics is revealing the human genome to be much more malleable, creative, and adaptive than we ever imagined. For example, the idea of "horizontal gene transfer" among bacteria—a process in which organisms share parts of their genetic material with other organisms around them—has recently gained some ground among scientists. The neo-Darwinian synthesis remains foundational, but the scientific story of our evolving universe is unfolding much faster than the public's capacity to fully digest the changes.

However, the evolutionary worldview I envision is not built by science alone. And so I intend to pursue a broader inquiry into the influence of evolutionary thinking on all aspects of culture. Again, this book is first and foremost about evolution as a powerful and significant idea rather than evolution as a science. The difference is critical. Evolution as an idea transcends the more limited, gene-centered perspective that has come to dominate public discourse. As the book unfolds, I will explore things like the evolution of technology, the evolution of cooperation, the evolution of consciousness, the evolution of worldviews, the evolution of information, the evolution of values, and the evolution of spirituality and religion. I believe these are legitimate and important ways to speak about evolution, and indeed critical if we are to adequately understand our life and our world. But that doesn't mean they are first and foremost scientific approaches, or that they will always be consonant with current biology. These ideas are stretching and pushing the very limits of our understanding of this important concept.

Before we continue, I want to acknowledge that there are those who are fearful of evolution being used as a context through which to interpret human life and culture. The most common opposition is the one voiced by religious communities who associate evolutionary thinking purely with atheism and materialism. I have addressed these concerns already and I will continue to do so in the pages that follow. The other common objection is reputational. Like a high school kid who gets a bad reputation for hanging with the wrong crowd, evolution is rejected for its dubious historical associations. Unfortunately, it is true that evolution has been used in the past as a tool to justify the most disreputable social philosophies, such as Social Darwinism in the nineteenth century and the eugenics movement of the early twentieth century. Some may despair that I have already identified Marx as being inspired by evolution. After all, look at the historical horrors wrought by Marxism! Others manage to draw a direct line from the evolutionary thinking of Hegel and Darwin to Hitler's fascism. Evolutionary speculations outside of the science labs are dangerous, they would tell us, leading to delusional notions like "some cultures are superior to others" or "there is an inevitable direction to human history." Put those two conclusions together, so the thinking goes, and you have the perfect petri dish for all kinds of fascistic and totalitarian abuse. Certainly, we can see abundant evidence for that in the twentieth century.

In the wrong hands, all powerful ideas have the potential for abuse—the more powerful, the more dangerous they can be. Evolution is without a doubt a potent idea, but I hope to show that many of the problems and abuses referred to above are no more intrinsic to evolutionary thinking than fanaticism is to religious thinking or nihilism to scientific thinking. The failures above are failures of immaturity—the regrettable and often reprehensible growing pains of a culture coming to terms with an idea as explosive as evolution.

Despite these concerns, the promise inherent in an evolutionary worldview is truly tantalizing. In some respects, this promise is obvious and unavoidable; in others, it is subtle and hardly recognized. It is obvious in the sense that new insights into the science of evolution will inevitably help to revolutionize technology in ways we barely foresee. As science probes deeper into the mysteries of the last fourteen billion years and begins to more accurately understand the way in which nature crafts her handiwork, the technologies that result will surely be game-changers.

However, this is not really a book about the progress of science and technology; it is a book about the progress of meaning. And so much of human meaning today comes from the way in which we interpret science, the conclusions we draw about the scientifically revealed universe. Therefore, part of what I will be exploring is how our interpretation of evolutionary science is evolving in ways that will have a significant effect on how we make meaning in the coming decades and centuries.

Breakthroughs in our understanding of cultural evolution also promise rewarding insights. While some remain unconvinced that evolutionary ideas can even be applied at a cultural level, new researchers, social scientists, and philosophers are working to deconstruct the very nature of human culture and the relationship between individual and social development. They are seeking the principles and patterns that inform the trajectory of cultural evolution. Such information could provide a more sophisticated understanding of how and why societies progress (or regress) in the long parade of history. Imagine for a moment if we were able to identify some of the key leverage points that help facilitate and encourage positive cultural development. In a world in which multiple civilizations seem to be in a time of great flux, understanding such principles would surely be a great boon to the overall health of our ever-smaller global village.

At the subtler end of the spectrum is the potential of an evolutionary worldview to serve as a new cosmology, one that provides an authentic meeting point between science and spirit. Evolution, in this respect, has a unique capacity to be a source of spiritual fulfillment, of authentic meaning and purpose, renewing our faith in the possibilities of the future and inspiring us to reach for those higher potentials, individually and collectively. This is perhaps the most profound promise of an evolutionary worldview, with implications as far-reaching as human aspiration itself.

EVOLUTION: A BROADER VIEW

Today, our postmodern world is restless, suspicious of big ideas yet hungry for meaning—searching for a new worldview, one that can guide us through the turbulent waters of a new millennium. Evolution is the latest contender to step into that cultural vacuum. And even some mainstream thinkers are beginning to notice. "While we postmoderns say we detest all-explaining narratives," writes *New York Times* columnist David Brooks, "in fact, a newish grand narrative has crept upon us willy-nilly and is now all around. Once the Bible shaped all conversation, then Marx, then Freud, but today Darwin is everywhere." Brooks is certainly right when he points to the tremendous influence that evolutionary ideas are having on our culture more than one hundred and fifty years after Darwin published *On the Origin of Species*. Even when I turn on the television to watch the NBA playoffs, I see the latest Gatorade commercial: "If you want a revolution, the only solution: evolve." I wonder if Darwin would be happy or bemused if he were still alive to hear such words accompanying images of great athletes scoring, sweating, celebrating, and hydrating.

It must be noted that Brooks is referring to an evolutionary perspective that is primarily defined by attempts to apply Darwinian ideas to social sciences and cultural studies. Indeed, it seems like a book comes out every week that features a scholar applying evolutionary thinking to some new field of study. Why do we have religious feelings? Why was Lady Macbeth so hungry for power? Is the impulse to go to war bred into our DNA? Why do we like snazzy technology? Such questions are examined by exploring how the origins of these attitudes and proclivities can be traced to the conditions of life in our evolutionary past.

For example, according to this perspective, we can analyze the ins and outs of sex, love, marriage, and all the cultural extras that go along with them, according to how they each played their role in the evolutionary drama of our ancestors. We get jealous because infidelity threatens our need to pass on our genes to the next generation. Women are attracted to powerful men because once upon a time in the tundra, their survival depended on aligning themselves and their children with a strong provider and protector. Men struggle with monogamy because they are genetically predisposed to sow their seed far and wide. And on it goes. According to this way of thinking, we tend to overindulge in fatty, sugary, and salty foods because those desires best served our survival and reproductive needs thousands of years ago. We band together in close-knit groups because the tight bonds of such relationships would have conferred survival advantages to our Stone Age predecessors. And the latest popular assertion is that humans have an evolutionary predilection for religiosity. Faith may be part of our genetic and cultural programming, goes this argument, because of how shared beliefs and rituals facilitate the creation of highly loyal social groups that helped us survive the slings and arrows of life among the Flintstones.

Such research is certainly illuminating, and much of it credible. Unfortunately, proponents of exciting new fields sometimes overreach, and evolutionary psychology, as this approach is often called, is no exception. Just as the field of Freudian psychology, in its heyday, tended to explain everything under the sun in terms of the Oedipal complex and childhood experiences, suddenly it's in vogue to explain all of human behavior by appealing to Darwinian processes. In the more extreme expressions of this trend, evolution has come to represent the de-facto reference point for seemingly all human motives—as if religion, morality, altruism, love, evil, marriage, infidelity, music, poetry, and so on can all be traced solely to the industrious activity of selfish genes. While there are certainly more nuanced versions of this story, it has, regrettably, caused many to adopt a suspicious attitude toward evolutionary thinking on the grounds that it carries with it a dangerous reductionism that circumscribes rather than expands our insight into human life and culture.

The problem here, however, is not evolution; it is a specific and narrow definition of the term. If we're going to use evolution as a context in which to examine human nature and human culture, I suggest that we are much better off expanding the way we think about this important idea. Evolution, as an idea, transcends biology. It is better thought of as a broad set of principles and patterns that generate novelty, change, and development over time. In some sense, this book is an exploration of those principles and patterns, as well as an examination of how they are illuminating and transforming different fields of human endeavor. Exclusively using Darwinian mechanisms as the interpretive lens for understanding the evolution of consciousness (psychology) and culture (sociology) may be insightful and interesting. But like using a magnifying glass to look at the *Mona Lisa*, sooner or later we are going to realize that there is a lot more to the big picture.

Nevertheless, the seed is sown, and the momentum for a larger evolutionary embrace of life is growing as new generations of thinkers consider the physical, biological, and cultural story of evolution as a new metanarrative. And it's easy to understand why. Evolution is at its heart an inquiry into who and what we are as a species. Simply put, it is our origin story. Evolution tells us where we've come from; it explains the historical roots and context of our very existence. Origin stories or creation myths have formed the basis of cultural worldviews and religions throughout history. They are one of the most universal expressions of our search for meaning. In African tribal legends, we find sky serpents whose coils formed hills and valleys and brought forth stars and planets, and white giants who vomited out the sun, moon, stars, and all living things. China's ancient *Tao Te Ching* describes the "nameless void" that gave rise to existence, split into yin and yang, and became Mother of the Ten Thousand Things. Norse mythology tells us that fire and ice met in a "yawning gap" and gave birth to life. In these and countless other tales, from the bizarre to the profound, humanity's creation myths attempt to answer what scholar Robert Godwin poetically calls "the perennial questions that have puzzled human beings ever since they became capable of puzzlement: how did we come to exist, what is the point of existing, and is there any escape from what appears to be an absurdly brief slice of existence between two dark slabs of eternity?"

By exploring where we've come from, what we're made of, and the factors that have molded and shaped us, evolutionary thinking has much to say about what makes us human, about what potentials and possibilities inform not only our brains and biology but also our consciousness and creative capacities. Are we mindless mechanisms, created accidentally by a meaningless material process? Are we special children of God, created in his image for a temporary sojourn in his earthly garden? Or are we something else? That "something else" is as yet undefined, but it is what I will be exploring in the pages that follow. There is a significant diversity among the emerging views I will be presenting, but that in itself is one of those principles of how evolution works—in nature, in culture, and even in the development of knowledge. As cosmologist Brian Swimme puts it, "You'll have an explosion of animal forms at the birth of a species—an explosion of *diversity*, this incredible chaotic explosion of possibility—and then the universe sort of winnows out the more exotic shapes and enfolds them into forms that are more enduring. Diversity is a great way in which the universe explores its future." As I explore the diversity of evolutionary perspectives that are vying for prominence in this new worldview, I hope I am contributing to that exploration—of the universe's future, and our own.

CHAPTER TWO

Breaking the Spell
of Solidity

*In laying hands upon the sacred ark of absolute permanency,
in treating the forms that had been regarded as types of fixity
and perfection as originating and passing away, the* Origin of
Species *introduced a mode of thinking that in the end was bound
to transform the logic of knowledge, and hence the treatment of
morals, politics, and religion.*

—*John Dewey,* The Influence of Darwin on Philosophy

Worldview" is a popular term these days, and for good reason. The word comes from the German *Weltanschauung*, and is used in common parlance to signify the framework we use to interpret the world around us. In our postmodern world, we have come to recognize just how important these interpretive frameworks are in shaping our perspectives and the perspectives of others. Some of this is a natural result of globalization and our increasing proximity to peoples and cultures that see the world through dramatically different eyes. "Why do they hate us?" asked President Bush in the week following 9/11—a question echoed on numerous magazine covers and newspaper headlines around the country and on the lips of stunned Americans who had never even considered such a thing as a worldview before. America was forced to come to terms with the fact that there were other people who see the world through a completely different lens—a lens so different that what to us was unthinkable, to them became horribly necessary. Even within our own diverse country, it is becoming increasingly clear that the differences between us are not just surface political or religious affiliations, they are more fundamental differences in how we interpret and experience the world around us and within us.

We may think that we simply have a direct perception of the world, but in fact, every perception is filtered through our particular perspective, as becomes clear in moments when we are confronted with someone whose perspective is dramatically different from our own. As philosopher Ken Wilber puts it, "What our awareness delivers to us is set in cultural contexts and many other kinds of contexts that cause an interpretation and a construction of our perceptions before they even reach our awareness. So what we call real or what we think of as given is actually *constructed*—it's part of a worldview."

There is actually a place where they study amorphous things like worldviews—the Center Leo Apostel, a research institute affiliated with the Free University of Brussels. They define a worldview in the following way:

> A world view is a system of co-ordinates or a frame of reference in which everything presented to us by our diverse experiences can be placed. It is a symbolic system of representation that allows us to integrate everything we know about the world and ourselves into a global picture, one that illuminates reality as it is presented to us within a certain culture.

A worldview is not so much a value; it is the very conglomeration of conclusions about the world that will determine what kind of values we hold. It is not just a collection of thoughts or ideas; it is the very structures of the psyche that will help determine what kind of thoughts or ideas we will have. Worldviews are like invisible scaffolding in our consciousness, deep conclusions about the nature of life that help shape how we relate to just about everything else around us. As the Christian scholar N. T. Wright explains, worldviews "are like the foundations of a house: vital, but invisible. They are that *through* which, not *at* which, a society or an individual normally looks."

We don't choose worldviews the way we choose a set of clothes or decide on our musical preferences. Worldviews are built on the cognitive and psychological architecture of the self and are heavily informed by the culture in which we live. They are not simply tastes we pick and choose at the cultural buffet line, conscious augmentations to our personalities—a dose of conservatism here, a helping of religion there, a plate of social liberalism on the side. No, worldviews are bound up in the very development of the self in the context of any given culture. We don't have them; for the most part, they have us. They are deep structures that determine the very way we make meaning in the closeted capacities of our own consciousness.

We might say that worldviews help us make sense out of the experience of being alive; they are, in other worlds, epistemological. They are also ontological, meaning that they speak to the way in which we understand the fundamental nature of being itself. But before you start thinking that worldviews are abstract ideas, let me disabuse you of that notion. Growing up in a small town on the edge of the Bible Belt, one learns at an early age that worldviews are frighteningly practical. To a teenager, they determine critical things like who can dance at parties, who is OK with premarital sex, and who thinks both things are an act of Satanic possession. They inform who goes to your church, or if one goes to church at all. They answer questions pertaining to race and sexuality. They help establish how one views ethics and morals. They delineate the possibilities inherent in manhood and womanhood. They liberate and constrain, give confidence and are cause for doubt. They are, we might say, the true tectonic plates of our global culture, and their movements determine a great deal about the direction and development of our society over time.

A TOUCHSTONE PROPOSITION

So where do we start in defining a new evolutionary worldview when its contours are as yet unformed? We can begin by asking: what is such a worldview based on? Indeed, at the center of any worldview is a core conviction or set of convictions about the nature of what is real, true, and important. So while worldviews may very well be complex psychosocial beasts, they are also, paradoxically, simple. I don't mean that they are simplistic, but rather that they are built on simple foundations, deep convictions that set the parameters and define the terms on which we construct self and culture. A worldview might express itself through individuals in hundreds of thousands of ways, but each of those expressions will carry with it the character of that foundational conviction.

Philosopher William H. Halverson suggests that "at the center of every worldview is what might be called the 'touchstone proposition' of that worldview, a proposition that is held to be *the* fundamental truth about reality and serves as a criterion to determine which other propositions may or may not count as candidates for belief." For example, we might say that the touchstone proposition of a modernist scientific worldview is that the universe is objectively comprehensible using rational inquiry and scientific methodology—a conviction that informs its interpretations of every dimension of life, from religion to art to economics.

I believe that the touchstone proposition for an evolutionary worldview is best captured in a passage by Teilhard de Chardin. It is from the first paragraphs of his classic collection of essays, *The Future of Man*, and sums up not only the basic distinction that lies at the heart of an evolutionary worldview but the essential spirit of it as well:

> The conflict dates from the day when one man, flying in the face of appearance, perceived that the forces of nature are no more unalterably fixed in their orbits than the stars themselves, but that their serene arrangement around us depicts the flow of a tremendous tide—the day on which a first voice rang out, crying to Mankind peacefully slumbering on the raft of earth, "We are moving! We are going forward!" . . .
>
> It is a pleasant and dramatic spectacle, that of Mankind divided to its very depths into two irrevocably opposed camps—one looking toward the horizon and proclaiming with all its newfound faith, "We are moving," and the other, without shifting its position, obstinately maintaining, "Nothing changes. We are not moving at all."

We are moving. I keep coming back to that fundamental insight, and appreciating how profound it really is. The things that we think are fixed, static, unchanging, and permanent are in fact *moving*. In so many areas of human knowledge, we are discovering that reality is part of a vast process of change and development. Like geologists discovering plate tectonics for the first time, we are beginning to look out at this extremely solid, seemingly permanent world that feels so stable underfoot, and intuit a radical truth: nothing is what it seems. *We are moving.* We are *going somewhere.* It is a slow but irrevocable revelation, dawning on our awareness. Our bedrock assumptions, it tells us, our most basic instincts about life and the universe are in error. Whatever solid ground we are standing on is itself in motion. We are not just being; we are *becoming*. That's part of the revelatory power of an evolutionary worldview. It's an ontology of becoming. We do not just exist *in* this universe; we are caught up in its forward movement, intrinsic to its forward intention, defined by its drift forward in time.

So many of the critical insights that people have come to in relationship to evolution boil down, in essence, to this one simple proposition. But even for those of us who accept and appreciate the basic principle of evolution, I don't think the extent of its influence has penetrated very deeply into our conscious awareness.

Several of my Californian friends have described the profoundly disconcerting experience of being in an earthquake, suddenly finding that the ground was moving under them for the first time. Nothing can prepare you for that moment, they told me. Psychologically, it is hard to take in, because something you considered so unquestionably solid— the earth underneath your feet—is *moving*. That which you considered absolutely fixed and stationary, is in fact not stable at all. And that seismic shift can create tremendous shock waves, not just in the surrounding landscape but in the fabric of the human character, because we have spent a lifetime unquestioningly trusting that solid foundation.

In a sense, there's an earthquake happening in human culture right now, and there has been for the past couple of hundred years. We have been captivated by the spell of solidity, the fallacy of fixity, the illusion of immobility, the semblance of stasis, but the evolution revolution is starting to break that spell. We are realizing that we are, in fact, not standing on solid ground. But neither are we simply adrift in a meaningless universe. *We are moving*. We are part and parcel of a vast process of becoming. The very structures that make up our own consciousness and culture are not the same as they were one thousand years ago, and in one thousand years they will be substantially different from how they are today.

We see this insight in so many fields of study. Most obvious, perhaps, is biology. Only a few hundred years ago we related to biological species as if they were more or less permanent. Species didn't change; they didn't evolve; they didn't go extinct—that's how we saw the biosphere. But Darwin's work demonstrated beyond any shadow of a doubt that the entire biological world is not fixed or static. Life is not just being; it is becoming.

The same is true at a cosmological level. Physicists used to think that we existed in what they called a "steady-state" cosmos—no beginning, no end. Suddenly, almost overnight, our picture has changed. The universe had a beginning. And it seems that it will someday have an ending. We are not drifting aimlessly in an immense cosmic sea but seem to be part of a vast developing process, the parameters of which we are barely beginning to grasp.

Similar revelations are dawning in our understanding of human culture. We now know that the socioeconomic systems and structures of society are not fixed or God-given or a result of unchangeable, eternal truths about human nature. They are adaptive structures that change and evolve over time. We can look back and begin to fathom the extraordinary transitions that have occurred in human culture in the last hundreds of thousands of years and see that the illusion of a solid, unchanging, static "way that human beings are" is up for question as never before.

This insight also has spilled over into psychology. In the nineteenth century, James Mark Baldwin, who was a pioneer in evolutionary theory, began to point out that even the categories of our psychology aren't fixed. He noticed that children are actually passing through developmental stages on their journey to adulthood. This was a radical idea at the time: the very structures of our psyche go through critical changes over the course of our lives. Today, we are realizing that not only do children change and develop but adults can as well. There is little if anything final or fixed about adult psychology.

Or consider neuroscience. We once thought the brain was static, fixed, and relatively unchanging; now we're discovering it to be more plastic and malleable than we ever dreamed. "Neuroplasticity" is a word on the lips of many these days, and for good reason. The spell of solidity is cracking in neuroscience and we are realizing that even the very gray matter so intrinsic to our sense of self is anything but permanent. It is developing in relationship to many factors, not the least of which are our own choices. In discipline after discipline, stasis is losing the battle to movement, process, change, and contingency.

Moreover, it's not just the world *out there* that is moving; it's also the world *in here*. It's not just the objects you see that are moving and evolving; it's also the subject, the perceptive faculty itself. The part of you that sees, listens, interprets, and responds is also not static or solid but rather is fluid, changing, caught up in a developmental process, non-separate from this fundamental characteristic of our evolving cosmos.

These are insights that go to the core of what it means to be human. They affect our own internal world, our deepest values, beliefs, and convictions. From the foundations of the self to the edges of the cosmos, we are starting to recognize that we are part of and, indeed, inseparable from this process. *We are moving too.* In fact, some might say that we are movement itself. In so many ways, this fundamental insight is emerging everywhere. One of my favorite metaphors for this shift of perspective comes from Henri Bergson:

> Life in general is mobility itself; particular manifestations of life accept this mobility reluctantly, and constantly lag behind. It is always going ahead; they want to mark time. Evolution in general would fain go on in a straight line; each special evolution is a kind of circle. Like eddies of dust raised by the wind as it passes, the living turn upon themselves, borne up by the great blast of life. They are therefore relatively stable, and counterfeit immobility so well that we treat each of them as a *thing* rather than as a *progress*, forgetting that the very permanence of their form is

only the outline of a movement. At times, however, in a fleeting vision, the invisible breath that bears them is materialized before our eyes. . . . allow[ing] us a glimpse of the fact that the living being is above all a thoroughfare, and that the essence of life is in the movement by which life is transmitted.

I love this metaphor because I'm from Oklahoma, and in the dry, hot days of my childhood summers I remember seeing what we called "dust devils" rising up from recently plowed fields. These were tornadoes of dust, sometimes small and fleeting, sometimes hundreds of feet high and imposing, borne up by the great gusts of Oklahoma wind, helter-skelter tempests racing across the plains in a doomed and desperate search for permanence. In those "fleeting visions" that Bergson described, we can sometimes see, for a moment, that even the most seemingly solid forms in the world around us—our environment, our cultural institutions, our bodies, our minds—are in fact like that dust, held in place only by the power of the invisible current of evolution that carries us. They are not permanent. They are more motion than matter. *The very permanence of their form is only the outline of a movement.*

Alfred North Whitehead, the great English Evolutionary and process philosopher, also spoke to this point when he suggested that reality is made up not of bits and pieces of matter but of momentary "occasions" of experience that fall and flow into one another and create the sense of reality and time, just as cascading hydrogen and oxygen molecules create the actuality of a river. He called our failure to recognize this movement, our tendency to turn flow into fixity, "the fallacy of misplaced concreteness."

Today, that fallacy is slowly crumbling. The spell of solidity is breaking. But we have not yet embraced the implications. "Permanence has fled," writes scholar Craig Eisendrath, "but it has left a world conceived as process, contingency, and possibility. The more we understand it, the more it increases in wonder. It is a world which we can help create, or lose, by our own actions." As we start to incorporate this new way of thinking and understanding the world into our consciousness, it will profoundly affect not only how we see the cosmos but also how we see our own lives. Unlike a physical earthquake, which leaves one feeling out of control, breaking the spell of solidity, while disconcerting, is ultimately quite liberating. No longer the victims of unchangeable circumstances, trapped in a pre-given universe, we find ourselves released into a vast, open-ended process—one that is malleable, changeable, subject to uncertainty and chance, perhaps, but also, in small but not insignificant ways, responsive to our choices and actions.

The pioneering men and women whom I have called Evolutionaries express the touchstone proposition of this new worldview in diverse voices. But what they share is the fundamental recognition and embrace of its truth. Evolutionaries are those who have woken up, looked around, and realized: *We are moving*. And rather than bury their heads back in the sands of seeming stasis, they are ready to pick up the paddles and help steer that raft that Teilhard envisioned toward a more positive future.

As the fog of fixity lifts, we are finding ourselves much more than observers and witnesses to life's grand unfolding drama. We are influential actors, newly aware of the immense tides that are shaping the world within and without, just becoming cognizant of our own freedom—and immense responsibility.

CHAPTER THREE

What Is an Evolutionary?

It is as if man had been suddenly appointed managing director of
the biggest business of all, the business of evolution—appointed
without being asked if he wanted it, and without proper warning
and preparation. What is more, he can't refuse the job. Whether
he wants to or not, whether he is conscious of what he is doing
or not, he is in point of fact determining the future direction of
evolution on this earth. That is his inescapable destiny, and the
sooner he realizes it and starts believing in it, the better for all
concerned. . . .

—*Julian Huxley, "Transhumanism"*

I f you wish to converse with me," the French philosopher Voltaire
is said to have remarked, "define your terms." Voltaire's wisdom
applies doubly when introducing what is essentially a new term like
"Evolutionary" into a discourse. And so I would like to take this
chapter to explain and expand on what I mean by this term, which is
beginning to be used by greater numbers across our culture today.
Perhaps the closest word to "Evolutionary" in today's parlance is the
term "evolutionist," a word commonly associated with evolutionary
theory in academic circles. "Evolutionist" is defined in dictionar-

ies as a person who is an "adherent to the theory of evolution." As suggested by that distinction, it is a term that has traditionally been associated with a person who strongly believes in and is influenced by the scientific theory of evolution. It is a term often contrasted with "creationist" or "biblical literalist" or other various Darwinian dissenters who proliferate on the reactionary edges of modernity.

Clearly there is much overlap between Evolutionaries and evolutionists. But as I implied in chapter 1, I intend for Evolutionary to mean more than that. Evolutionary is a play on the word "revolutionary," and I mean it to convey something of the revolutionary nature of evolution as an idea. Evolutionaries *are* revolutionaries, with all the personal and philosophical commitment that word implies. They are not merely curious bystanders to the evolutionary process, passive believers in the established sciences of evolution, though all certainly value those insights. They are committed activists and advocates—often passionate ones—for the importance of evolution at a cultural level. They are positive agents of change who subscribe to the underappreciated truth that evolution, comprehensively understood, *implicates* the individual. Indeed, an Evolutionary is someone who has internalized evolution, who appreciates it not only intellectually but also viscerally. Evolutionaries recognize the vast process we are embedded within but also the urgent need for our own culture to evolve and for each of us to play a positive role in that outcome.

With that in mind, I would like to outline three critical characteristics common to Evolutionaries. This is hardly an exhaustive list, but I hope it manages to capture the essential spirit of this designation. First, Evolutionaries are cross-disciplinary generalists. Second, Evolutionaries are developing the capacity to cognize the vast timescales of our evolutionary history. Third, Evolutionaries embody a new spirit of optimism. I will explore each of these characteristics and their significance in the pages that follow.

IN DEFENSE OF THE GENERALIST

This is not a world built for generalists. It is a world built for special-ists. What's valued intellectually is specialty knowledge—expertise on the mechanics of eukaryotic cells or the chemistry of black holes or the life cycles of ant colonies. Even within individual disciplines, the drumbeat of specialization takes precedence over broader sys-tems of knowledge. It's not enough to be a physicist; one is a par-ticle physicist or a quantum-loop theorist or a string theorist. It's not enough to be a historian; one is an expert on Renaissance social customs or South Asian political dynamics in the eighteenth centu-ry. Indeed, the degree of specialization in our collective knowledge base is both stunning in its depth and detail and frightening in its increasing fragmentation.

"Most educated people at the beginning of the twenty-first cen-tury consider themselves to be specialists," writes Craig Eisendrath. "Yet what is needed for the task of understanding our culture's evo-lution, and of framing a new cultural paradigm, is the generalist's capacity to look at culture's many dimensions and to put together ideas from disparate sources."

Evolutionaries are generalists for this very reason. Readers will notice that the critical insights that populate these pages are a re-sult of thinking as a generalist must think—with a passionate but broad curiosity that fans out across culture and sees connections, patterns, transitions, and trends where others only see discrete facts and details. An Evolutionary must be able to look at the movements of nature, culture, and cosmos as a whole, yet without denying the infinite detail that surrounds us.

If one reads the books written by many of today's evolution-ary thinkers, this is one characteristic that immediately stands out. Whatever their fields of expertise, most are incredibly well-informed generalists. They move from one field to another with ease and sometimes brilliance. They are unafraid to risk the wrath of the specialists and take research from one field and apply it to another. They are interpreters par excellence—synthesizers, holistically inclined pattern-recognizers. They mine today's incredible knowledge base for insights and help make sense of the enormous confusion that the information revolution hath wrought. In doing so, they serve a great function. They help explain our place in the scheme of things.

Of course, there are times when such thinking can go very wrong—for example, when well-intentioned but ill-informed people take difficult concepts from a complex field such as quantum physics and draw overly facile conclusions about how they apply to spirituality and life. Bookstores are filled with such ill-conceived problem children of the science-and-spirit relationship. And it's not just spirituality. *New York Times* columnist Paul Krugman has used the term "biobabble" to describe a similar misapplication of biological principles to economic systems. Moreover, even if our thinking is clear and our intentions genuine, it is always hard to satisfy the specialists' criteria, to avoid stepping on toes in fields that are not one's primary expertise. I am sure that I, and many of those featured in these pages, will invite such criticisms. But that should not deter us from appreciating the importance of this missing function.

In recent decades there has been a growing sense that the critical role that a generalist plays in society is being forgotten, with dangerous consequences for our culture. In discipline after discipline, experts have raised concerns that our knowledge base has privileged depth and detail over breadth and context. As Eisendrath points out, one result of this increasing fragmentation of knowledge is that there is no one left "to speak for the culture as a whole."

So who is responsible for this overwhelming fecundity of fragmentation? Scapegoats abound, but the person most frequently cited is a six-hundred-year-old philosopher—René Descartes. Truth be told, Descartes is guilty only of articulating an important breakthrough that characterized the changes occurring in his own time period. It was Descartes who announced the radical split between subject and object that the world has been struggling to come to terms with ever since. He placed the thinking, rational self, the *subjective* self, in a position distinctly separate from the rest of the universe—the *objective* world. *I think, therefore I am,* he famously declared—*Cogito ergo sum*. It was the foundation of the Cartesian revolution. In that one great statement, human beings announced a further extrication of their own consciousness from its primordial embeddedness in the natural world. And not just the natural world. Our consciousness was also becoming free of its immersion in the social life of the group or collective, a process Canadian philosopher Charles Taylor called the "great disembedding." In Descartes's declaration, the modern self, we might say, found its liberation and autonomy. Obviously, the statement itself did not catalyze this change, but Descartes's words and subsequent philosophy helped establish a new and powerful way of thinking that empowered the changes that were occurring in consciousness and culture of the day. The "I" was becoming free of the "it" and of the "we." Human beings could begin to see *objectively* as never before, allowing us to gaze with new eyes upon nature as an external object of curiosity and fascination, viewed from the dispassionate stance of the separate observer.

The result was a cultural revolution that led to the European Enlightenment. Out of that great split of man from nature came the modern world and all of its many wonders, the first and foremost of which was the modern individual, newly autonomous and free to define him- or herself on his or her own terms.

In a sense, we could say that Descartes broke the world in two, metaphysically cleaved it down the center, and the reverberations are still felt today. Indeed, it was as if that fracturing, once under way, had a momentum of its own, and in the wake of that one great split came a thousand smaller cracks. Whole systems of knowledge began to separate out from one another, finding their own freedom and relevance, released at last from the unifying strictures of a once-dominant religious worldview. Our religions, with their antiquated belief systems and values, could no longer contain or make sense of this multidimensional, diverse world that was bursting the boundaries of a premodern intellectual edifice. Science and philosophy broke away from religion, fracturing into their own separate domains, which allowed them to develop free of the superstitions of the medieval church.

With this development, the reign of religion as the grand unifier and dominator of culture thankfully was over. New fields of study began to arise as the human spirit was liberated to investigate the natural world as never before. Like unstable particles unable to cohere, the sciences subdivided and subdivided into the mass of compartmentalized specializations that mystify unwary college applicants today.

And so the task in our time has changed. We have gained all the power of specialization, recognized the necessity of reductionism, practiced the art of slicing and dicing reality into smaller and smaller revelations, but now we must set a new course. We have so much information but so little context. We have so much knowledge but somehow lack a larger framework in which to understand it. We are data rich and meaning poor. It takes me all of ten seconds on Google to find the infant mortality rate of Chad in 2003, and yet we have seemingly no clue as to how and why some cultures evolve in healthy ways and others descend into anarchy. We have mapped the marvelous complexity of the human genome and yet stand by helpless as kids wander our streets as dropouts and junkies, undeveloped throwaways of the wealthiest culture in history. We may be on the verge of unlocking the very secrets of life and longevity, and yet millions of people have so despaired at our capacity to positively influence the evolution of culture that they have decided the only way forward is for the Earth to suffer a near apocalypse or, as some believe, undergo a miraculous global awakening. Evolutionaries sense that the world is fragmented, and that we must embrace our role in jump-starting the process of reintegration.

Evolution, by its very nature, helps us to integrate our thinking. It transcends the neat structures of disciplines mapped out on the university campus and encourages us to lift our eyes to patterns and trends that break the boundaries of compartmentalization. It compels us to think in bigger ways about life, time, and history, until finally we find ourselves staring at contexts so fundamental that they can temporarily break the hold of the mind's incessant fascination with particulates of experience and reveal completely new perspectives on existence. Perhaps that is why Hegel, one of the original evolutionary philosophers, when asked "What is truth?" replied with the slightly flippant but no less profound answer: "Nothing in particular."

Arguably, the sciences have been most responsive to this need for more integrative thinking. For example, the renowned Santa Fe Institute was formed in the late 1980s to facilitate a new kind of cross-disciplinary dialogue. One of the original associates of the Institute, the Nobel laureate physicist Murray Gell-Mann, noted that while the process of specialization was necessary and even desirable,

> [T]here is also a growing need for specialization to be supplemented by integration. The reason is that no complex, nonlinear system can be adequately described by dividing it up into subsystems or into various aspects, defined beforehand. If those subsystems or those aspects, all in strong interaction with one another, are studied separately, even with great care, the results, when put together, do not give a useful picture of the whole. In that sense, there is profound truth in the old adage, "The whole is more than the sum of its parts."

This study of complex systems that Gell-Mann is referring to is usually called complexity theory or systems science. As it turns out, the principles that govern the behavior of complex systems transcend any particular discipline. The same principles that govern the functioning of the stock market might also help shed light on the behavior of a hive of honeybees or the growth patterns of a megacity. These principles cannot be contained by the neat human-created categories that separate the physics department from the sociology department down the hall. They apply generally. The evolutionary principles explored in this book are similar. They cannot be confined to biology or sociology or theology. Specialists risk a kind of myopic tunnel vision. They often cannot see cross-disciplinary evolutionary principles, much less apply them.

This approach reverses centuries-old predilections in scientific disciplines that argue that truth is best found by breaking wholes down into parts and those parts further into their respective parts, and so on. There is nothing inherently wrong with this kind of scientific approach. In fact, the fruits of such thinking are all around us—from increasingly powerful smart phones to life-extending medical breakthroughs. And yet we have come to recognize that there is so much more to reality than can be captured by this approach. "Reductionism alone is not adequate," writes Stuart Kauffman, "either as a way of doing science or as a way of understanding reality." Even many of the most committed so-called reductionists recognize that such a perspective, adhered to religiously, slices off whole segments of reality. Amid the proliferation of the parts, we so easily lose the whole. We may know the physical makeup of the puzzle, and how every last piece is shaped, but until we put them together, we can have no true sense of the actual picture.

Science is hardly alone in its attempts to reach beyond fragmentation and specialization toward a more unified approach to knowledge. Philosophy also has managed to leap forward toward a more "integral reality," largely through the work of individuals like philosopher Ken Wilber, whose work we will examine in chapter 11.

And religion, likewise, has a growing number of believers in a less siloed approach to matters of spirit.

But despite these initiatives and many more like them, integration is still a road less traveled. The generalist remains a rare breed, and the evolutionary generalist even more so. There are few who have the capacity or inclination to speak for "culture as a whole." Yet there is little question that our future lies in this direction. As author James N. Gardner writes, in what I think is one of the most salient and inspiring descriptions of precisely this kind of integrative attitude toward knowledge:

> The overlapping domains of science, religion, and philosophy should be regarded as virtual rain forests of cross-pollinating ideas—precious reserves of endlessly fecund memes that are the raw ingredients of consciousness itself in all its diverse manifestations. The messy science/religion/philosophy interface should be treasured as an incredibly fruitful cornucopia of creative ideas—a constantly coevolving cultural triple helix of interacting ideas and beliefs that is, by far, the most precious of all the manifold treasures yielded by our history of cultural evolution on Earth.

Being an evolutionary generalist is more than simply being a pluralist—one who makes space for multiple perspectives and points of view. In fact, there is evidence, coming from a variety of sources, that integrative, cross-disciplinary thinking may not just be the latest and greatest idea of the cognoscenti but an actual higher mental function that represents a further step in the evolution of consciousness itself. In other words, it may be an evolutionary adaptation to the challenges presented by our globalizing, ever-complexifying society. At least, that is the testimony of individuals like twentieth-century German philosopher Jean Gebser (whose work we will explore in chapter 9). He was convinced that a new consciousness was dawning in human life, one that he distinguished from the "mental/rational" consciousness that had characterized the modern era. He called this new consciousness "integral," and wrote that it was characterized by what he called an "aperspectival" quality, meaning that it contained a way of seeing reality that transcended the segmentation and fragmentation of the mental/rational world-view. "Our concern is with integrality and ultimately with the whole," he wrote.

One person whom Gebser pointed to as an example of this new integral consciousness was the great twentieth-century Indian philosopher-sage Sri Aurobindo. In his masterpiece, *The Life Divine*, Aurobindo outlined in remarkable detail the gradual ascension by which human cognition moves from one stage of mind to the next higher stage. The level he called "higher mind" he describes as the capacity to take in knowledge by intuitively perceiving it as an integral whole, an all-at-once perception of multiple ideas grasped simultaneously as a unified truth. The best analogy I can think of would be an orchestra. We experience great music as a coherent whole but remain aware of the extraordinary melodies and wonderfully complex harmonies that contribute separately to that singular delight.

Such descriptions also call to mind another early twentieth-century icon, James Joyce, who used the term "epiphanies" in his stories to describe the same sort of revelation. Joyce's characters would undergo what he called a "simple sudden synthesis" of insight and understanding that contained the qualities of wholeness and integrity. And in our own time, Ken Wilber has preferred the term "vision-logic" to describe this curious mixture of visionary revelation combined with conceptual and logical analysis that seems characteristic of this new mental, yet transmental, capacity referred to by each of these figures.

We can see, even in these brief examples, that for some theorists, evolution is not just happening in the external world. The very faculties we use to perceive the world are themselves caught up in the evolutionary process. These theorists suggest that the relatively limited capacities of *Homo sapiens sapiens* in the twenty-first century represent not some final end state of development or a completed picture of human possibility but merely one more stage in a cosmic drama that has taken us from energy to matter to life to mind and now seeks higher and higher potentials. They suggest that the immense challenges of our globalizing world are themselves catalyzing and calling forth evolutionary potentials in human consciousness, which will allow us to begin to make deeper sense of the immense complexities of our wonderfully diverse but painfully fragmented age. No, they aren't teaching this in Kansas schoolrooms or creationist colleges, but neither is it common at Harvard. However, if we are to form a more perfect union of our fragmented world in the days to come, it is a perspective worth considering.

DEEP THOUGHTS IN DEEP TIME

If there is a complementary trend in evolutionary thinking to the return of the generalists, with their integrated approach to knowledge, it is the impulse to look at reality through the lens of what I call *evolutionary time*. It is not dissimilar to the way evolutionary biologists must look at the species they study. Biologists understand that the limited time frame in which they see any given species is too small to reveal the true extent of the evolutionary changes that all species are undergoing. Except in rare cases, we cannot "see evolution" in our time scale. Dramatic and important evolutionary changes happen outside of our frame of reference, and therefore we must break the spell of "local time" upon our consciousness and lift our attention to much longer, more expansive time frames in order to envision the truth of evolutionary change and development.

After Darwin published *On the Origin of Species* in 1859, this challenge of understanding the time scales was one of the greatest obstacles to the acceptance of his theory of natural selection. People just couldn't get their minds around the amount of time required to make the process work. They had to infer. The same is true when thinking about evolutionary change in any context. *We must think in evolutionary time.* We must think with that unique kind of historical context.

Teilhard de Chardin suggested that the capacity to see in deep time is an emergent potential in the species. We are learning to perceive the vast epochs involved in the evolutionary dynamics that make up our bodies, and even our minds. And Teilhard used a fascinating analogy: There is a certain point in a child's development when he or she gains depth perception for the first time. Up until that point, everything in the baby's perceptual space is all organized on a flat plane, but at a certain point, the visual context deepens and objects begin to spread out in three dimensions. Teilhard compared our own emerging recognition of evolutionary time at this moment in history to a baby getting depth perception for the first time. We are just beginning to cognitively grasp the time context of our evolutionary emergence, just beginning to see in a new dimension. Time has been called the fourth dimension, so perhaps we're just beginning to develop the capacity to "see" in four dimensions.

"Just as we separate in space, we fix in time," wrote Henri Bergson, echoing precisely this point. "The intellect is not made to think *evolution*." Indeed, evolution has not yet equipped our brains to naturally think in such a way that we are able to deeply perceive the historical context of our own emergence. And yet miraculously, in a flash, we do see. The blinders of local time are removed and we suddenly grasp the extraordinary truth that we are ancient—related to all of life and connected to the history of the cosmos itself. We are not entities observing, as if from a distance the ongoing flow of time. No, we are windows into deep history itself, momentary formations of individuality composed not just of matter but of vast and primeval rivers of time.

It is almost as if a new form of spiritual intuition is dawning upon those with the inner eyesight to perceive it. Evolutionaries often report that this internal evolutionary timescape, while sometimes expressed powerfully in the written word, film, or other media, can also arise suddenly in consciousness, analogous to the flash of insight characteristic of a spiritual awakening. In fact, almost all of the individuals I interviewed for this book have had such an evolutionary awakening, though it has taken different forms. In many respects it is similar to what we think of as a spiritual or mystical experience. But while the experience may be filled with great meaning and spiritual significance, the content is less focused on Spirit or God, in a traditional sense, and more on evolution, process, and change. Awakening to a felt sense of the past and the future as much vaster than ever considered before, the individual feels connected to the developmental, in-process, unfolding nature of his or her own consciousness, of culture, of life, and even of the cosmos itself. The spell of solidity is broken deep in the recesses of the psyche and a new vision of an evolving world pours forth, an epiphany not just of unity and oneness but of movement and temporality.

THE RETURN OF OPTIMISM

A number of years ago, I began to notice that almost all of the Evolutionaries I encountered, all of the individuals who were inspired by the potential of using evolution as a context for understanding life and culture, had a third quality in common that made them stand out from their cultural milieu. In addition to being cross-disciplinary generalists and being able to think across evolutionary time, they also demonstrated a profound faith in and commitment to the future. They radiated a powerful optimism, one that stood out even more for being so counter to the overwhelming mood of the moment. Indeed, *Evolutionaries are deep optimists*. I'm not talking about a naïve optimism, a forced optimism, a superficial optimism, or even a hopeful optimism but an informed confidence in the knowledge that evolution is at work in the processes of consciousness and culture, and that we can place our own hands on the levers of those processes and make a positive impact. It is a subtle but powerful current of conviction that lifts the sails of the psyche and propels it forward into the future. Evolutionaries don't just believe that the future can be better than the past; somehow they know it—like a great leader knows that she can make a difference; like a great athlete knows that he can compete and win.

I would suggest that the unique flavor of this evolutionary optimism cannot be attributed to a mere personal feeling, inspiration, or belief. It runs deeper than that. Evolutionaries evince a confidence that is different from the brashness and bluster that flows out of the personal ego. It carries with it a conviction that reaches beyond any quality found only within the boundaries of the personality. And they transmit that confidence to others. We tend to transmit to others how we feel about life at a fundamental level. When one spends time with a great mystic or saint, there is a quality to the personality that is recognizable, whatever the particular tradition of that individual or belief system—a quality of ease, of deep peace, and of transcendent being that we experience in the company of those whose source of confidence lies far deeper than the individual psyche. The same is true of this evolutionary optimism. It arises from a direct perception of the possibility of evolutionary development and connects us to drives and impulses that are neither personal nor cultural, drives that some feel are connected to creative forces at work in the evolving universe. It is as if the essence of the process itself—its creativity, dynamism, and forward movement—comes alive in the personality of those who have embraced an evolutionary worldview.

It is important to note here that the evolutionary optimism I am speaking about does not equate to a conviction in an inevitable positive outcome, or a belief in a miraculous "shift" that is just about to happen. We see this kind of thinking all too often in spiritual-but-not-religious circles—whether it be a Mayan prophecy, the Harmonic Convergence, or some sort of "Earth Change" that will pave the way to the future. Such ideas are often held by individuals with the best of motives, who look out at a world of climate change, terrorism, corruption, overpopulation, and financial disaster, where billions live in poverty, and conclude that things are not getting better at all. Or if they are, they aren't improving fast enough. And then they pray, hope, meditate—for some event; some change of consciousness; some immanent convergence, emergence, or resurgence of love, light, peace, and compassion to deliver us from the darkness and ignorance that has a hold on our collective soul. And too often, they invoke the term "evolution" to describe this shift in consciousness.

Such thinking has nothing to do with evolution as I understand it. In fact, I would suggest that it is not a faith in evolution that leads one to embrace such naïve or exaggerated hopes but, in fact, a lack of faith. It is an insufficient appreciation of the power of evolution and a failure to understand how it works, at a cultural level, that leads some to start reaching for super-historical forces to emerge and save the day. When we begin to appreciate the true dimensions of the vast evolutionary process that we are a part of, our optimism becomes grounded in the slow but demonstrable reality of actual development.

When I was a boy, I would watch the great tennis players of my era, Björn Borg and John McEnroe, vie for the Grand Slam titles. And I would imagine what it might be like to be able to play like that and compete at such a high level. That vision I held in my mind was important to my development as a player. But ultimately, what was more inspiring and invigorating was to experience myself learning actual new skills, however far they may have been from those of my heroes—to see the reality of self-development and build confidence in the fact that I could transform myself into a better player through my own efforts. Such an experience temporarily breaks the spell of solidity, at least in relationship to our personal capacities. When we remove the illusion of immobility, it's like breaking a dam in our consciousness. We start to see the world around us in new ways, experience new and liberating possibilities, and see more directly the underlying momentum that is part of the process of human nature and human culture. We start to see how we can actually *choose* to develop and mature, individually and collectively—whether it's on the tennis court or in much more important areas of human society. *That* is the source of evolutionary optimism.

There is nothing wrong with great visions of possibility. We need them, as long as they're not crazy and unrealistic. They inspire us and give us direction and focus. But what truly uplifts and invigorates us is to participate in actual development, and in doing so to appreciate how this development is connected to the larger historical flow of evolution since the beginning of human culture. When our eyes open up to the reality of evolution and we can look back and see not merely thousands and thousands of years of survival and endurance but centuries and centuries of hard-won progress, we will stop hoping for miracles. We will embrace a deeply optimistic vision of the future, one that empowers us to embrace the challenging but ultimately much more rewarding work of contributing to a process that transcends our own lives and that, miraculously, we can affect with our own actions.

Indeed, the grounded optimism and positivity that lights up the hearts and minds of Evolutionaries makes quite a statement in our cynical, meaning-starved culture. I hope you will recognize it in the pages ahead, and perhaps begin to feel it yourself as we dance across disciplines and plumb the depths of cultural and cosmic time. It is a conviction not only in the fact of evolution but in the wholesomeness of the evolutionary process, despite the suffering, conflict, and chaos it inevitably entails. In the hearts of these Evolutionaries, the future is already bright.

PART II
REINTERPRETING
SCIENCE

CHAPTER FOUR

Cooperation: A Sociable Cosmos

Ecological communities are not simply gladiator fields dominated by deadly competition; they are networks of complex interactions, of independent self-interests that require mutual adjustment and accommodation with respect to both the other co-inhabitants and the dynamics of the local ecosystem. The necessity for competition is only one half of a duality, the other half of which includes many opportunities for mutually beneficial co-operation.

—*Peter A. Corning,* The Synergism Hypothesis

I've always loved watching Steve Nash play basketball. And it's not because he's the tallest, the best shooter, the fastest dribbler, or the most effective rebounder. No, Nash, a relatively small, self-effacing Canadian, is a delight to watch precisely because he's not the perfect specimen of a basketball player. In a one-on-one matchup, he'd lose to Kobe Bryant, LeBron James, or Dwyane Wade every time. But there is a reason why Nash's Phoenix Suns have been perennial contenders. Basketball is a team sport. The advantage goes to the collective that thrives, not the individual who strives. And Nash's amazing gift—his adaptive advantage, we might say—is his capacity to transform the disparate individuals on his team into a cohesive unity, to turn five capable players into a sort of superorganism of basketball that can accomplish things five individuals would never dream of. That's what earned him the NBA's Most Valuable Player Award two years in a row. And that brings us, surprisingly, to the work of biologist Lynn Margulis.

In 1967, Margulis, a young unknown biologist who just happened to be married to a young astrophysicist named Carl Sagan, published a landmark paper. She argued that millions of years ago, single-celled organisms began to work together, resulting in the development of an entirely new life form—the eukaryote (the first nucleated cell)—that became the basis of all advanced life on the planet. Margulis's work on this new theory, which she called "symbiogenesis," was a watershed event in the development of evolutionary biology, helping to shift not only the scientific conversation but also the cultural conversation surrounding evolution from a focus on *competition* to a new appreciation of *cooperation*. She helped show how cooperation among organisms—in this case, bacteria—could be a major driver in the evolutionary process. Today that essential insight is being used to understand everything from ancient tribal dynamics to the ongoing march of economic globalization. Just as Sagan helped us appreciate how insights into our cosmic heritage could shed light on the meaning of human life and culture, Margulis has helped us recognize that there are revelations aplenty to be found in the "microcosmos," as she has dubbed the world of the infinitesimally small. In fact, she has scolded the scientific community for narrowly focusing on animals, organisms that, when all is taken into account, have made a much more recent appearance in evolution's long story.

"Animals are very tardy on the evolutionary scene, and they give us little real insight into the major sources of evolution's creativity," writes Margulis. "It's as if you wrote a four-volume tome supposedly on world history but beginning in the year 1800 at Fort Dearborn and the founding of Chicago. You might be entirely correct about the nineteenth-century transformation of Fort Dearborn into a thriving lakeside metropolis, but it would hardly be world history."

The breakthrough work on symbiogenesis represents the classic story of an unknown outsider fighting against the scientific establishment to have her work accepted. In 1967, Margulis published her original work in a paper called "On the Origin of Mitosing [Eukaryotic] Cells." Doesn't exactly sound like light summer reading material, but in the world of lab coats and microscopes, this paper, like her subsequent book, was a must-read, reframing the origins of multicellular life on the planet. She argued that one of the most critical advances in evolution's past, the formation of the nucleated cell, which is essential for all higher life forms (sort of the biological equivalent of the discovery of the wheel) was made possible by the *cooperation* of early bacteria:

> My major thrust is how different bacteria form consortia that, under ecological pressures, associate and undergo metabolic and genetic change such that their tightly integrated communities result in individuality at a more complex level of organization. The case in point is the origin of nucleated (protoctist, animal, fungal, and plant) cells from bacteria.

This landmark paper was not exactly embraced by the scientific powers that be. Remember, this is the same scientific community that once described the first eons of Earth's history as "three billion years of non-events." They were fascinated by fruit flies and searching for fossils from millions of years ago; they were less interested in the evolutionary relevance of protists and prokaryotes from billions of years ago. After rejection by fifteen scientific journals, Margulis finally got it published, and little by little her "radical" idea won over the scientific mainstream. But the whole process left her bitter about the reigning evolutionary consensus that was more focused on competition between selfish genes than cooperation among symbiotic organisms. Outspoken and provocative by nature, she once referred to the mainstream neo-Darwinian evolutionists as "a minor twentieth-century religious sect."

I first began to appreciate the work of Margulis through the work of another evolutionary biologist—the self-described "bio-philosopher" Elisabet Sahtouris. Sahtouris and I met at a panel discussion on science and spirit during the 2000 State of the World Forum in New York City. She immediately stood out as someone who had an unusual knack for cross-disciplinary communication. She could wax eloquent about the evolutionary dynamics of the microbial world without resorting to overly technical language and capture the subtleties of science in ways that made them powerfully relevant to contemporary concerns. Yes, Margulis's 1966 paper may have heralded a major scientific breakthrough, but "On the Origin of Mitosing [Eukaryotic] Cells" doesn't exactly get the heart racing—at least, not outside the subscription list of the *Journal of Theoretical Biology*. When Sahtouris took her own shot at describing the creation of the first eukaryotic cells, the result was considerably more, well, engaging. As she explained it:

> The tiny archaebacteria, with their specialized lifestyles and technologies, then created the most dramatic event to occur in Earth's evolution since their own initial appearance out of the Earth's mineral crust. The nucleated cell—an entirely new life-form about a thousand times larger than an individual bacterium—formed, as the bacteria took on divisions of labor and donated part of their unique genomes to the new cell's nucleus. Thus, the nucleated cell—the only kind of cell other than bacterial ever to evolve on Earth—represents a higher unity than the bacteria achieved after eons of tension and hostilities,

as they engaged in successful negotiations and cooperative evolution. This process—whereby tension and hostilities between
individuals lead to negotiations and then ultimately to cooperation as a greater unity—is the basic evolutionary process of all
life forms on our planet, as I see it.

In the hands of Sahtouris, bacterial evolution had much more
than scientific relevance; it had dramatic flair. Here we have tension and negotiation, new technologies and old hostilities. We even
have bacterial "lifestyles," whatever they might be. We have, in
short, the essential characteristics of a story—and a good story, as
we shall see—a true triumph-over-tragedy tale of epic dimensions.
Sahtouris somehow has a knack for infusing the activities of the
microbial world with a kind of contemporary relevance, helping to
give us a new appreciation for our original ancestors and the billions
of years in which bacteria ruled the planet. She writes:

> Before our new wave of knowledge about our single-celled
> ancestors—bacteria and protists, or nucleated cells—the bulk of
> evolution was as murky a prehistory as the three million years of
> human existence prior to what we call the Stone Age. Now, quite
> suddenly, we are unveiling a surprisingly cosmopolitan ancient
> (and modern) microworld. Discovering the urban lifestyles of
> bacteria with all their technologies—from skyscrapers to com-
> pass and electric motor, from solar energy devices to polyester,
> and even to a world wide web of information exchange—is an
> amazing journey.

The analogy between the bacterial and human worlds only goes
so far, but it is fair to say that scientists are continually surprised at
just how sophisticated and complex bacterial colonies are proving
to be. Their collective intelligence, meaning their inherent capacity
to turn aggregate action into intelligent behavior, is off the charts.

For example, researchers are discovering, much to the dismay of our health-care system, just how resilient bacteria can be to antibiotics as they are constantly evolving, improving, and reinventing their gene pool. And each of Sahtouris's descriptions of "technologies"—electric motor, solar energy, polyester, etc.—while designed to be evocative rather than technically accurate, are also based on real capacities and functions within the biofilm communities that continue to impress and surprise researchers who inevitably underestimate the capacities contained in the microcosmos.

What makes this microscopic world so interesting for our inquiry into evolution is that, as unlikely as it may seem, there is a relationship between the dynamics at play in the rough-and-tumble world of globalization in the twenty-first century and the dynamics at play in Earth's prebiotic soup billions of years ago. This is one of the most interesting aspects of an evolutionary worldview—it allows us to move between the multiple levels and scales of the life process and see the same principles. And one of those principles is this: the spoils of evolution go not to the fastest or the smartest but to those who can find the best relationship between creative individuality and cooperative sociality. Among those ancient bacteria was, we might say, the Steve Nash of bacteria. And as we shall see, somewhere along the line that Most Valuable Protist and others like him helped turn a collection of competing bacteria into a more complex community of creativity and cooperation. We could call it the sweet spot of evolution, that perfect middle place between competition and cooperation that avoids blind self-interest on the one hand and uncreative groupthink on the other. It is the creative tension between the expression of individuality and needs of the collective, as our bacteria ancestors knew at some deep archetypal level almost two billion years ago. As Israeli scientist Eshel Ben-Jacob noted, "The aesthetic beauty of these [bacterial colonies] is striking evidence of an ongoing cooperation that enables . . . bacteria to achieve a proper balance of individuality and sociality as they battle for survival."

Evolution Cycle: Competition to Cooperation

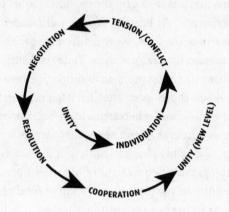

STRESS AND THE SINGLE BACTERIA

A few years ago at an event in Philadelphia, Elisabet Sahtouris quickly had me engrossed as she laid out her understanding of the complex dynamics of living systems. The evolutionary process goes through a seven-stage cycle, she explained, beginning with unity and then cycling through a series of stages laid out in the chart above.

Evolution, in this model, proceeds from *unity* to extraordinary *diversity* (individuality), which then leads to a competition for resources. That competition inevitably creates new *conflicts* and the period of conflict precedes a period of *negotiation* and eventually leads to a *resolution* of the resource problems, which in turn moves us back along the cycle to a place of *cooperation* and *unity*. Sahtouris's map is not so much a circle as an ascending spiral, such that the new character and quality of the unity that is discovered after cycling through all of these stages of development has a greater complexity, higher levels of cooperation and differentiation, and therefore, ultimately, greater consciousness.

This evolutionary cycle played out dramatically in one of the great environmental crises to grip the biosphere in its first few billion years of existence. As bacteria *differentiated* under the various selection pressures of the time, several different species with different functions roamed the prebiotic seas. Today, scientists work their way back to that primordial moment and imagine how one species may have, in its search for food, attached itself to something called a glucose gradient—a border between a more nutrient-rich area and one less dense in nutrients. Here, along these fertile borders in the ancestral soup, was food aplenty. Think of it as bacteria taking up residence in the prebiotic equivalent of a border town, an ancient hub where all kinds of travelers (in the form of food resources and energy flows) were likely to pass through. And it is here that we come across another principle of evolution that was active in the earliest forms of life but that applies equally to the postmodern age as to the Protozoic. *Evolution happens at the edges.* Evolution happens on the borders, the boundaries, the in-between zones.

This is true whether we are talking about nature or culture. It was the case in ancient glucose gradients that helped spur the creation of eukaryotic cells, as well as in the primordial mud between land and sea where scientists suggest that life first emerged. It's a principle that can be seen on the border between earth and space where the biosphere came into being, as well as at the edges of the Roman Empire where Christianity took root and flourished. It's why smart technology companies develop "Skunk Works"—small groups of committed, creative people who develop new innovations on the edges of their own institutions, free of the inhibitive, more conservative structures of normal corporate life. Evolution happens at the edges. Don't think of it as an absolute law but as a guiding principle that proves surprisingly true again and again and again. My favorite example is that of New Orleans in the nineteenth century. It was here, on the boundary between three or four cultures, that new innovations in music gave rise to the beginnings of jazz.

Let us return to Earth's earlier days, when glucose gradients provided a gathering point for single-celled bacteria to mix and mingle, work together, trade ideas, and even engage in a little experimental sex—swapping genes and energy. Somewhere along the way, a new innovation arose in the bacterial community, an early form of photosynthesis, as industrious bacteria developed a way to gain energy from exposure to ultraviolet light. But that would prove problematic, as the main waste product of this activity, the excretion of these photosynthetic creatures, was oxygen. Indeed, Margulis's work suggests a deadly *conflict* formed between various bacteria, a life-and-death struggle caused by what author Howard Bloom calls the first "toxic pollutant holocaust" in the Earth's atmosphere. It was akin to an ancient version of climate change. The photosynthetic bacteria's excretion of oxygen might seem a benevolent activity to us modern, oxygen-loving creatures, but not to the rest of the bacterial realm. The theory is that eventually such large amounts had amassed that a whole new layer of oxygen was formed around the Earth, creating the oxygen-rich chemical composition in the atmosphere that may sustain life today but was deadly to the ancient bacteria. Starvation ensued because bacteria were cut off from essential life-sustaining access to energy and food. Barbarism ran rampant, scientists speculate, and a bacteria-against-bacteria civil war unfolded over food and resources. But even as this great famine deepened, significant new adaptations began to emerge that would shift the conflict and move our one-celled ancestors further along on the evolutionary cycle. In fact, it would lead them to form the first eukaryotic cells, the breakthrough cells that became the critical foundation for more complex forms of life. Margulis describes the events in the following way.

How did the eukaryotic cell appear? Probably it was an invasion of predators, at the outset. It may have started when one sort of squirming bacterium invaded another—seeking food, of course. But certain invasions evolved into truces; associations once ferocious became benign. When swimming bacterial would-be invaders took up residence inside their sluggish hosts, this joining of forces created a new whole that was, in effect, far greater than the sum of its parts: faster swimmers capable of moving large numbers of genes evolved. Some of these newcomers were uniquely competent in the evolutionary struggle. Further bacterial associations were added on, as the modern cell evolved.

Here we have the next two stages of our cycle, *negotiation* and *cooperation*, leading eventually to the emergence of the final stage. The new *unity* in this case was represented by the original eukaryotic cells—small municipalities of working bacteria that were a thousand times larger than their one-celled ancestors. Once aggressive bacterial conflicts evolved into cooperative work associations as organisms took up residence in these new cells—independent bacteria now functioning in microbial partnerships, increasing access to energy and resources. We call them mitochondria. And so this particular circuit of the evolutionary cycle was complete, and a new social grouping was created that would change life on the planet forever. Sahtouris likes to riff on this theme, noting that the trajectory of evolution moves from one-celled creatures (bacteria), to multi-creatured cells (eukaryotic cells) to "multi-celled creatures" (humans), which may soon form new multi-creatured cells (global cooperatives?). Indeed, with the advent of globalization and the ever-growing need to come together in higher levels of organization at a planetary scale, it seems that we are repeating this cycle, working to form a new, multi-creatured global cell.

Sahtouris loves to draw an analogy between the "Great Oxygenation Event," as it's called in some science textbooks, and our current global climate crisis. But I was more struck by another point she made. Mark it down as yet another significant principle of evolution learned by studying nature's smaller creatures, although it's equally applicable to larger ones: *Stress creates evolution*. Certainly the emergence of eukaryotic cells is Exhibit A. After all, it was the ecological pressure of the atmospheric calamity that produced the kind of novelty needed for this breakthrough in the evolution of life. "Some of the greatest catastrophes in our planet's life history," Sahtouris concluded, "have spawned the greatest creativity."

I wanted to be sure of her words, and so I approached her after the talk to introduce myself. After some pleasantries, I asked her directly, "Do you really think stress creates evolution?" With a surprising intensity she looked at me and exclaimed, "Stress is the *only* thing that creates evolution." Now, she was talking about living systems in biology, but my mind was on a different class of living system—human beings. Stress, challenge, and adversity do indeed create evolution, not just in one-celled bacteria but potentially in the consciousness of the complex, multi-celled creatures that are currently in control of this beautiful blue planet that we (and billions and billions of smaller creatures) call home.

SELFISH GENES OR SOCIAL SCENES?

In order to appreciate just how significant the work of Margulis and others has been in presenting evolutionary theory in a new light, we need to go back and look at the way in which evolution has generally been represented in the culture at large. As evolutionary ideas began to infiltrate the sciences in the nineteenth century, people began to draw all kinds of conclusions about human nature based on the current state of evolutionary theory. They still do so today. This is natural and inevitable to some degree, but dangerous as well. Indeed, I have often been surprised in my own research to see just how closely some theories of human nature tend to follow the outlines of science. However, there is an important distinction between having one's philosophy of life *informed* by science and having it be *deter-*

mined by science. An evolutionary worldview embraces the findings of science, but it also gives human agency and free will its due. If evolutionary biology tells me that my nature, biologically speaking, is warlike and competitive, I can accept that truth and let it appropriately inform my thinking without in any way taking that to be the final word in the complex story of our human character. And when science evolves, as it must inevitably do, and lo and behold, it turns out that my nature, biologically speaking, is full of cooperation and altruism, I can let that, in turn, inform my thinking without letting it absolutely determine my worldview. In other words, science is an open-ended story, and any conclusions we draw based on it had better be tentative, temporary, and open-ended as well.

Alas, real life is not always so simple. When Jane Goodall, the famous primatologist, was working with chimpanzees in Africa, she initially thought that these primates, who share at least 95 percent of their genes with humans, were peaceful and cooperative. Chimpanzee society seemed to be free of the warlike behavior so common to human society. But looks can be deceiving. Goodall eventually observed that she was mistaken, that organized aggression is very much part of the social fabric in chimpanzee societies. When she began to report her findings, she received a great deal of resistance from her colleagues in the scientific community. She was even encouraged to suppress her data. The concern was not with the science but with the social consequences of such a discovery. It was a politically incorrect truth. Wouldn't the revelation that chimpanzees are warlike be a blow to our own efforts to transcend warfare? For some, the answer clearly was yes, because for them, the predilections of the chimpanzees seemed to represent an inherent truth about our genetic predispositions. And they were willing to sacrifice hard-earned scientific evidence to support their social agenda. Of course, if we live inside a reductionist worldview that tells us that human behavior is entirely driven by our genetic makeup and that we are bereft of free will and the power to influence our own future, then Goodall's discovery that our closest genetic cousins are not gentle pacifists may indeed be cause for despair. But such a conclusion is not inherently scientific.

"A little learning is a dangerous thing," the poet Alexander Pope famously noted. We would do well to heed his words when it comes to applying the scientific fads of the moment to the complexities of human nature. This was especially true in the nineteenth and early twentieth centuries, when intellectuals around Europe and America began to use the excitement surrounding Darwin and evolutionary theory to imagine how evolution might be applied to economics, social relations, and a host of other issues. There were the now-discredited claims of the eugenics movement (a term coined by Darwin's nephew), which promoted the idea of selective breeding and sterilization to achieve a more intelligent human race. Supporters of eugenics worried that the poor were outbreeding the rich, and that human civilization might be headed for an evolutionary disaster in which the lesser genes of the maladapted and unintelligent lower classes would somehow crowd out the clearly superior genes of the upper classes. Imagine a world in which the right to conceive children depended upon an IQ chart or one's income level, as if we could breed superior humans the way we breed faster racehorses.

Even a cursory reading of eugenics should have marked it as the Idea Most Likely to Be Abused of the twentieth century. But this danger was lost on a generation of early Western geneticists who gave their professional support to the notion. They called for the "self-direction of human evolution," a phrase we hear now in very different kinds of contexts. The term "social Darwinism" has been used to refer to exactly this kind of application, usually misguided, of Darwinian principles of survival and adaptation to the economic and social realities of life. In some quarters, it became a justification for inequality—if I'm rich and you're poor, it's just survival of the fittest—and helped contribute to regressive social attitudes and policies.

While our appreciation of the complexities involved in using evolutionary science as a tool for social policy has certainly improved in the intervening years, evolution itself still has a bad reputation. It gets labeled as promoting a dog-eat-dog version of the world, one that implicitly encourages the "red in tooth and claw" aspect of our natural heritage. "Since the origin of evolutionary biology," writes Stanford biologist Joan Roughgarden in her recent book *The Genial Gene: Deconstructing Darwinian Selfishness*, "Darwinism has been synonymous with competition and selfishness." Whether this reputation is deserved or not is a matter of some debate, but celebrated Oxford biologist Richard Dawkins certainly hasn't helped matters with his emphasis on the "selfish gene" as the defining characteristic of human nature. "We are survival mechanisms," Dawkins enthusiastically declares. "Robot machines, blindly programmed to preserve the selfish molecules known as genes." With statements like that, it doesn't take a genius or a geneticist to figure out what's wrong with the public conception of biological evolution. It's not the science; it's the marketing.

So are we competitive and selfish or cooperative and social? Certainly much has been gained in evolutionary biology by understanding the competitive, genetically driven character of our biological nature. And Dawkins's theory of the selfish gene is, no doubt, as philosopher Michael Ruse recently described it, one of the brilliant metaphors of the last century. Moreover, no one should suspect for a moment that our emerging understanding of cooperation has somehow smoothed the rough edges of nature and turned mammals into peace-loving team players. My cute and cuddly cat may have a deeply symbiotic relationship with her human family, but she has a bloodthirsty relationship with the local mouse and chipmunk population, and a fiercely competitive one with every other cat on the block. Nature is still, as Ken Wilber likes to say, one big restaurant, with everything eating everything else, and there is plenty of evidence that our biological heritage is full of competitive agents seeking to fulfill their own selfish ends. But in the view of Margulis, Sahtouris, Roughgarden, and others, too much has been made of the selfish drivers of human behavior. This new wave of science looks at the evolutionary process from stem to stern and sees marvelous example after marvelous example of cooperation and sociality in the service of evolution. Their mission is to free evolution from the taint of the selfish-gene metaphor and the resulting confusion about what it means to be human. In putting the emphasis on "social selection," the evolutionary focus shifts toward the survival not just of the fittest individuals but of the most effective social arrangements. The evolutionary advantage goes to those most capable of good teamwork and most willing to engage in the kind of cooperation that turns a collective of individuals into something more than the sum of its parts. The close historical relationship between evolution and selfish individualism is being decoupled, breaking apart an unhappy marriage that was always more hype than substance. Evolution is evolving, and so is the story we tell ourselves about life and what makes us human.

It's worth noting that this new story of cooperation transcends the realm of biology. It can be seen at work in the very movements of matter in the primordial cauldron of cosmic evolution. According to evolutionary theorist Howard Bloom, sociality was built into the original formation of life itself because carbon—the miracle molecule on which all of life is based—is uniquely structured to be able to "hook up" with other promiscuous elements in the periodic table.

So no matter where you look in this vast universe, from the lowliest bacteria to the smallest quark to the highest hominid, cooperation and sociality seems to be part of the picture. Once again, that does not mean that competition is dead or that we live in a touchy-feely Hallmark-card universe. Just ask Jane Goodall. Animals still kill, and blood still flows. Humans still do plenty of killing too, for that matter. We don't have to look far to find reminders that selfishness is indeed written in our genes. Yet somewhere in the midst of that picture there is an unmistakable shift in our understanding of the evolution of life and the cosmic order. We are just beginning to appreciate how the human urge toward friendship, toward connection and camaraderie, toward deep solidarity and true companionship, toward working together in more and more profound ways, has a real and demonstrable evolutionary precedent in the workings of matter and the organization of life itself. Without a doubt, science is evolving. Our view of human nature still has some catching up to do.

HUMANS, HIERARCHY, AND OTHER PROBLEMS THAT PLAGUE POST—SELFISH GENE SCIENCE

In the 1997 blockbuster movie *Men in Black*, the characters played by actors Will Smith and Tommy Lee Jones track down an interstellar fugitive on the streets of New York. This rogue alien is attempting to steal "the galaxy on Orion's belt" from a rival alien race. In a wonderful Hollywood twist, the galaxy, as it turns out, is not a vast macrocosmic formation in the far-distant sky but rather a marvelous microcosmic galactic world hidden away in the bauble on a cat's collar. During a pivotal scene, the pug dog alien, who is the source of this revelation, scowls at Will Smith and exclaims, "You humans, when are you going to learn that size doesn't matter?!"

It occurred to me somewhere in the midst of learning about the microcosmic world of bacteria that this might be a good motto for the evolutionary biologists who make the ultra-small so wonderfully fascinating for the rest of us. Indeed, the more we learn about our biological heritage and the remarkable evolutionary adaptations that have given rise to life as we know it, the more we appreciate just how connected we are to even the smallest and seemingly least among nature's inhabitants. It is a humbling realization. It is so easy, with all the accoutrements of modernity, to somehow see human life as separate from the natural world—to see our own psychology, sociology, and even physiology as existing independently, free-standing, disconnected from the multibillion-year evolutionary context that has given us life. Margulis, Sahtouris, and other like-minded scientists reconnect us with that history; they remind us that we were protists and prokaryotes and even quarks before we were anything with eyes or ears or thoughts. They undercut our anthropocentricism (and even mammal-centrism) and help us understand our inherent, inextricable connection with the dynamics of the biosphere. Cultural historian William Irwin Thompson described this sentiment powerfully when he noted, "So sensitized have I been by Margulis's work that I can now appreciate how bacteria have been treated like invisible serfs working in the fields while we humans dined in the manor house and talked about the evolution of consciousness as if it were only about the hominization of the primates and the emergence of the human brain."

I concur with Thompson's sentiment, and yet it does invite a question: Once we have properly chastised ourselves for our disconnected consideration of the privileged status of *Homo sapiens sapiens* and made appropriate adjustments to our worldview, where then does humanness fall in the evolutionary equation? What then is the hierarchy, if there is one at all, in which we can place both bacteria and human beings? How do we measure the difference in moral terms?

This question was brought home to me recently in a conversation one day at a conference in California. I had found myself unexpectedly at a lunch table with three well-known authors in the science-and-spirit field. I was quietly eating my salad when one of them asked me a question about my lunch preferences. I don't remember the details, only that I mentioned in my reply that I was a longtime vegetarian. This seemed to incite some concern.

"Why are you a vegetarian?"

"Well . . ." I began, only to be interrupted again.

"Don't you know that plants are conscious too? It's not like there is a big difference between that salad you're eating and animal meat."

"What?!" I exclaimed, momentarily stopped in my tracks. Many times I had heard people question vegetarianism on the grounds that it reflects an overly sensitive relationship to animal life, but never had anyone complained that I was doing just as much harm to plants!

I quickly recovered. "OK, but isn't there a difference between plants and animals when it comes to consciousness?" I asked. My lunch companions, however, were not to be deterred, explaining to me that all the latest research and experiments show that plants are conscious, and they experience pain. They seemed unanimously convinced that I was doing as much harm to my lettuce as I would to a chicken.

Increasingly bemused, I held my ground. "Whether one chooses to be a vegetarian or not, eating animals is different than eating plants. Animals have a brain, a face; they're more conscious."

But it was clear that in the eyes of my fellow diners, I had made an erroneous assumption, one full of ignorance and poor judgment. Only in California, I kept thinking to myself. Only in California would I have to defend myself for eating a salad! And for implying, however subtly, that plants somehow might not have the same onto-logical status as animals. Only in California would someone argue *against* vegetarianism on the grounds that plants are conscious too.

Later, when I told a colleague about the event, he reminded me of a statement made by the great Buddhist writer Alan Watts. When confronted with essentially the same question as to why he was willing to eat vegetables but not animal meat, given that both acts killed living things, he is said to have replied, "Because cows scream louder than carrots." It's a stark way to frame an important philosophical and moral issue, and one that all Evolutionaries must confront. Anytime we discover an important new insight into the natural world or resuscitate underappreciated parts of our biologi-cal (or cultural) heritage, there is a tendency to overstate the case. Simply because we see the tremendous contributions and previously unrecognized complexity of our bacterial legacy does not mean that we can therefore flatten every organism into a unidimensional moral soup. Did my friends in California really believe there was no difference at all between plants and animals? At the risk of being wrong, I actually doubt it. They were reacting to a reductionist worldview that denies the richness of sentience and subjective experience to the least among living creatures, even plants. But in their zeal to tear down a false hierarchy, they were in danger of unknowingly installing a frightening new absence of any hierarchy whatsoever.

We live in a time when false hierarchies are being questioned, in culture and in nature. This is one of the greatest gifts, and greatest potential dangers, of our pluralistic, egalitarian, postmodern worldview. One of those false hierarchies has been the assumption that microbial life was a bit player in the evolutionary drama. Every day, biologists are learning more about how wrong this assumption is. This realization has prompted some environmentalists and deep ecologists who are inspired by the work of Margulis and others like her to jump to the opposite conclusion—that there is somehow *no* essential difference between human life and other forms of life, including microbial life, thereby suggesting that they exist on the same moral plane.

"Evolution is no linear family tree but change in the single, multidimensional being that has grown now to cover the entire surface of Earth," writes Margulis. There are often spiritual overtones associated with this point of view, a celebration of the "oneness" and unity of life on Earth or Gaia, as our planet is fondly called. This perspective captures an important truth—that we are deeply connected to and nonseparate from the natural world. It is a vision that calls to mind the mystical insight and spiritual intuition of the Romantic tradition. Indeed, Margulis's words suggest a sort of nature mysticism and connect us to the visionary poets and writers who gave birth to modern environmentalism—individuals like Thoreau and Muir, who looked out on the wonders of nature and reminded the modern world that there is intrinsic value in the wilds, and that the human connection to nature is something we would do well to heed. After all, we are embedded in and dependent upon the biosphere, and failure to adequately understand or appreciate that insight has taken us down some dubious and dangerous paths in our stewardship of that "single multidimensional being."

However, there is a subtle yet significant difference between understanding humans as being *intrinsic to nature* and understanding humans as being *intrinsic to a natural evolutionary process*. If our attention is on the process, then what we care about is not only the health of the biosphere but also the health of the larger evolutionary process through which the biosphere gave birth to all the wonders of nature—and us. Indeed, Margulis's words bring home an important ecological fact, but they bypass an equally important evolutionary truth: that the biosphere is not simply a static and stationary Garden of Eden that we appreciate as if it were in suspended animation but a rich, creative cauldron of evolutionary ingenuity that has situated human beings in a unique position. Our human intelligence, our unique capacity to think and reflect, is part of nature too, and we dare not underestimate its worth. I have often been shocked at how many environmentalists who care passionately about the human impact on the natural world somehow forget that our many uniquely human capacities that distinguish us from the rest of the natural world *are also part of nature*.

I do not mean to defend the environmental record of my fellow humans. But I do strongly mean to suggest that a worldview that merely reintegrates humans as one equal part among the billions of inhabitants of the biosphere will never be enough. We cannot embrace nature and deny evolution. And evolution, by its very nature, implies hierarchy. Indeed, I firmly believe that we can elevate our appreciation for all of the many wonders of nature without simultaneously denigrating the evolutionary advance that human life represents. We are the progressive edge of evolution as far as we know, the creative dynamics of the Earth come alive in human form, and if we want to be able to make the kind of moral distinctions that we need to make in the twenty-first century, we must learn to appreciate that truth, and the responsibility, of that unique position.

If you're still on the fence about this matter of hierarchy, I'll leave you with a final example. A few years ago, many of my fellow editors and I were shocked when a respected spiritual teacher suggested that despite the fact that 9/11 was a bad day for humans, it was a great day for bacteria. After all, from the point of view of the bacteria that got to feed on those bodies, it certainly wasn't a tragedy, so who are we to judge what's right or wrong? This kind of conclusion is the inevitable result of a world in which everything is flattened out into a false oneness, when you make no hierarchical distinctions between lettuce and lambs, between humans and bacteria, and then follow the moral logic to its natural ends. I hope the absurdity of this example is self-evident. But many today at the leading edges of culture flirt with views not entirely dissimilar, views that question the worth of the human experiment altogether. In the name of combating a false anthropocentrism they embrace a nonsensical egalitarianism. In the most extreme cases, they at least imply that there is no difference between the rights of bacteria and the rights of humans. But we need more than an intuition that this assertion is wrong; we need a worldview that explains why. Admittedly, we have not had an adequate worldview in which the distinctions between bacteria, plants, animals, and humans, can be clearly elucidated. And science alone cannot provide that.

At that California lunch table, the moral issues may seem inconsequential. But they loom large in the increasing concern over the planet's ecological health and the human environmental impact. In a globalizing world, where the circles of our interdependence seem to be growing bigger every day, how are we to negotiate the moral, spiritual, and economic lives of seven billion human beings? How do we deal with the fact that many of those individuals are living with beliefs and worldviews that are incompatible with many of the others, not to mention those who have adopted lifestyles that are incompatible with the biosphere and the other forms of life that share our planet? There is simply nothing in our cultural history that adequately prepares us for the evolutionary challenges we now face, challenges that will demand more robust and sophisticated answers to the question of how human beings are actually related to the natural world out of which we have emerged.

Given these unprecedented realities, I think we can safely say that the bar has been raised when it comes to cooperation. We are facing a moral, spiritual, and interpersonal challenge that will test and transcend everything evolution has ever learned about how to make a community—large or small; bacterial, mammalian, or human—work. Our success, and even our survival, may depend on how we are able to walk the delicate line between cooperation and self-interest as the cycle of evolution takes a new turn, embracing the extraordinary interconnected diversity of our global community.

CHAPTER FIVE

Directionality: The Road to Somewhere

The drama of life is a cumulatively transformative process in which something of the utmost importance is taking place, even if analytical science cannot see it.

—*John Haught*, Making Sense of Evolution

S o, is evolution going somewhere?" I asked the man across the restaurant table from me on a balmy spring night in Tucson, Arizona. The sun had just gone down, ending the last day of the "Toward a Science of Consciousness" conference sponsored by the Center for Consciousness Studies at the University of Arizona. I had spent several days in Tucson as a journalist and researcher, curious as to the latest theories and activity in this relatively new field. The term "consciousness studies" may seem like something one is more likely to encounter at an Indian ashram than a major university, but increasingly science is turning its attention to perhaps the greatest mystery of evolution—consciousness itself.

Every couple of years, hundreds of pioneering academics, maverick researchers, and even a few conventional types, make the pilgrimage to Tucson to speak their minds and hear the latest about such unusual topics as "Use of Mathematical Physics to Model Neural Correlates of Brain Activity in Perception and Consciousness" or "Games Brains Play: Neurological Disturbances of Self and Identity" or "Retroactive Modulation of Subjective Intentions: Philosophy, Science and Cyborgs." One of those who had made the journey was a familiar face—John Stewart, an Australian evolutionary theorist and author of the book *Evolution's Arrow: The Direction of Evolution and the Future of Humanity*. I had attended his panel in the afternoon and listened to his interesting talk on "The Potential Future Development of Consciousness."

Stewart and I had first met at a conference years earlier and kept up a correspondence since. With a soft-spoken, easygoing Australian personality that meshes surprisingly well with a brilliant intellect, Stewart is the kind of person who can turn an impromptu dinner at a local restaurant into a long memorable evening of intellectual insight and stimulating conversation. And if that dinner just happens to take place on an outdoor patio in the unnaturally perfect Tucson spring weather, and is accompanied by a particularly nice Australian cabernet sauvignon—well, who am I to complain?

The topic of our conversation was directionality, or to use a more philosophical term, "teleology." For those not familiar with that term, it is an important and controversial one in the study of evolution. *Telos*, in Greek, means "end" or "purpose." So a teleological view, in relationship to evolution, is one that sees the process as having a particular purpose or direction—sees it as *going somewhere*, as opposed to just randomly unfolding. For some, the idea of teleology implies not only a direction but a specific, predictable end; for others, it means that the process has a clear directionality. It is this latter sense of the term that I use in this chapter.

The issue is critical as we outline the framework for a new worldview. The prevailing currents of evolutionary science have long been suspicious of the idea that one can glean any significance at all from the trajectory of cosmic history, and certainly not from the long march of life from simple bacteria to walking, talking, thinking mammals. The concern is that any sort of identifiable, recognizable directional trajectory in evolution inevitably smacks of purpose, purpose smacks of intelligence, and legitimizing ideas like purpose and intelligence will open the doors of science to all manner of unwelcome nonscientific speculations. This suspicion on the part of science dovetails nicely with the parallel conclusion in the humanities that the notion of progressive evolution in human cultural history is also a dangerous and unsound idea, one that should be, and largely has been, banished from the world of academia.

The concern, of course, with either of these positions is: are they true? Are they the best interpretation of the facts as we find them today? After all, if we're interested in the deeper dimensions of evolution, then directionality, purpose, intelligence, progress, and all the potential meaning that accompanies these terms are at the very core of our concern.

"Much of the hostility to the idea that evolution has a progressive trajectory goes back to the mid-[twentieth] century, when what's called the neo-Darwinian synthesis was just being formed," Stewart explained to me as we waited for dinner, already deep in conversation. "Beginning in the 1940s, major figures in evolutionary theory, including Julian Huxley and Theodore Dobhzansky, decided that evolution needed to be respectable, and therefore science-based evolution had better not dabble in speculative big-picture stuff. Even though many of these founders actually had progressionist perspectives themselves, they suppressed these approaches in order to build the professional standing of evolutionary biology within mainstream, reductionist science."

It is perhaps hard to fully appreciate from today's perspective just how much cultural pressure there was in the middle of the last century to rid evolutionary theory of its tendency to attract a teleological aura. Indeed, ever since Darwin (and in truth, even before him), evolution was being embraced by those who wanted to find much more than blind selection in the workings of nature's laws. By the 1940s, Darwin's breakthrough had already been used by people as diverse as American philosopher John Dewey, French philosopher Henri Bergson, and German theorist Rudolf Steiner as a means to not only reenvision biology but to rethink cultural history, sociology, philosophy, spirituality, education, and even theology. Some of this theorizing was intriguing, some of it profound, and some of it questionable at best, but all of it was stirring emotions, excitement, and controversy like few subjects can.

These speculations began to directly clash with the larger tides of twentieth-century culture. Two world wars had considerably dampened enthusiasm for any kind of teleological view of history. For many, directionality smacked of Marxism, and the West was at war with the dark results of that cultural experiment. Moreover, Hitler's fascism had been built upon the idea that one race of human beings had a special destiny, that they were intellectually and culturally superior to the rest of us and represented a higher step on the evolutionary ladder. Perhaps, many suggested, we should throw out big, speculative, meaning-laden ideas of history altogether. They were dangerous. Forget about historical trends, patterns, stages, or directionality. Embracing such ideas became seen as risky and morally hazardous, too closely associated with many of these darker forces in human history. And this was understandable. As British paleobiologist Simon Conway Morris writes, "That biology can be co-opted for agendas, if not ideologies, that promise an ever-more-perfect future, albeit across piles of corpses, is evident from the lunacies adopted by totalitarian states."

Evolutionary theory, its proponents felt, needed to free itself from such associations, and the best way to do so was to stick to the narrow confines of careful and concrete science. "The prestige of evolutionary research," wrote the famous biologist Ernst Mayr in 1948, "has suffered in the past because of too much philosophy and speculation." To the increasingly skeptical modern mind, confronted with an ever-more-fragmented world, the idealistic narratives of the evolutionist philosophers seemed to lack a sufficient empirical basis. The editors and peer reviewers of the important journal *Evolution* took care to eliminate any such language from its pages as evolutionary science embarked upon a period of consolidating its academic legitimacy.

Ironically, as Stewart explained to me, many of the great leaders in the field continued to publish on ideas of directionality and progress, still captivated (as Evolutionaries tend to be) by the teleological implications of evolutionary science. But in a strange sort of double life, they confined their speculations to popular books and eliminated such "philosophizing" from their professional endeavors. Scottish physiologist J. S. Haldane wryly remarked that "teleology is a mistress without whom no biologist can live, but with whom none wishes to be seen in public."

So despite brief bursts of infidelity on the part of the biologists, the overall trend of the last half-century was established, and those with a bias toward seeing direction, purpose, and progression in evolution seemed fewer and fewer, and their voices in the culture at large grew softer and softer. And that wasn't just true in the world of biology. Across the spectrum of the humanities, the postwar intellectual climate saw a retreat from reading directionality and its close cousins, purpose and progress, into the tea leaves of cultural history.

It's striking how many of the great champions of evolutionary progress came of age before World War I. This only underscores the change in western society that occurred between 1914 and 1945. Prior to Europe's twin conflagrations there had been a general belief in the forward movement of human culture. Those were the days of globalization's first incarnation, and while concerns about industrialization and modernity were also active, it was still possible to unabashedly believe in the future, to embrace the word "progress" without irony or without immediately qualifying it with a string of caveats and justifications. But all that had changed by 1945. The unguarded optimism that once flowed freely from modernity had finally run dry. Enthusiasm for progress in history seemed as dead as the millions laid to waste on the battlefields of Europe and Asia, and our cultural thought leaders took a collective step back to assess the validity of such notions in a world of atomic weaponry, genocide, totalitarianism, and environmental destruction.

Historian Massimo Salvadori writes in *Progress: Can We Do Without It?*, "The twentieth century was a great burial ground for ideas and bodies. . . . In this vast cemetery, the idea of Progress . . . was laid to rest both by those who had consciously rejected it and by those who had first undergone its moulding influence and then gravely deformed it." Directionality and determinism were rejected, uncertainty embraced. Belief in modernity's promise was replaced with a marked lack of belief, not just in the traditional gods of myth and magic but in the modern deities of technology and progress. We were slowly becoming *postmodern*. And for postmoderns then and now, directionality and progress are dangerous ideas, potential gateways to illusions of cultural superiority, environmental disaster, and economic domination. They are best relegated to the ideological dustbin of history.

Biologist David Sloan Wilson captures this concern in his *Evolution for Everyone*, in which he describes how a young graduate attempting to apply the insights of evolutionary theory to other fields ran into a wall of opposition based on many of these negative presumptions when he tried to discuss his newfound interest with his professors and peers:

> [He] quickly learned that when [he] spoke of human behavior, psychology and culture in evolutionary terms, their minds churned through an instant and unconscious process of translation, and they heard "Hitler", "Galton," "Spencer," "IQ differ-

ences," "holocaust," "racial phrenology," "forced sterilization," "genetic determinism," "Darwinian fundamentalism," and "disciplinary imperialism."

Wilson is not exaggerating. In the humanities, evolution is mistrusted as the facilitator of so much that was wrong with modernity—unchecked industrialization, cultural superiority, historical determinism, and the kind of reductionism that interprets all human motivation through the lens of Darwinian mechanisms. In academia, evolution is a science, nothing more, and any talk of philosophy, spirit, directionality, purpose, meaning, or grand overarching theories of life and history is heavily discouraged. At the furthest end of this spectrum, scientists such as the late Stephen Jay Gould saw progress as "a noxious, culturally embedded, untestable, nonoperational, intractable idea that must be replaced if we wish to understand the patterns of history."

One may be forgiven for thinking this to be the last word on the subject, that directionality might end up being a casualty of the evolution wars. Some have wished it so. From every corner, it seems, evolution as a larger progressive event, grand idea, or a great organizing narrative for life was rejected, ignored, or hemmed in to narrow boundaries, even as evolution as a narrow and specific science continued to flower and develop.

Today, we live in the world molded by this cultural legacy. But at the edges, much is shifting, churning, and changing. Stewart, my dinner companion that spring evening in Tucson, is one of a new generation of theorists who have thrown off the fears of the twentieth century and taken a fresh look at evolution with new eyes and new appreciation for its powerful teleological characteristics. Does evolution have a direction? Was human-level intelligence the inevitable, or at least a likely, result of the evolutionary process? And, most interestingly, if evolution has a direction, where is it headed? These questions are alive again in the intellectual currents of culture. Simon Conway Morris, a celebrated scientist who made his career studying the Burgess shale fossils, published *Life's Solution: Inevitable Humans in a Lonely Universe* in 2005, arguing that biological evolution has a direction, and that humans, or something essentially like them, were indeed a likely result of the process.

Morris suggests that life's tendency is to find the same evolutionary solutions despite disparate circumstances. "Rerun the tape of life as often as you like, and the end result will be much the same," he writes. For example, the eye was independently created many times, suggesting that even given varying natural conditions, eyes are ultimately a good solution to evolution's needs. Add up all such "convergent" solutions and it's fair to propose, according to Morris, that something looking and acting like a human was going to be the inevitable result. He even suggests that evolution on other planets might reasonably be expected to produce intelligent creatures not entirely alien to the two-legged walking, talking apes that are found here on planet Earth.

Robert Wright, author of *The Moral Animal*, *Nonzero*, and *The Evolution of God*, has taken up a similar banner, arguing to some acclaim that not just biological but also social evolution has a direction. Despite the twists and turns of cultural development, he argues, a clear path can be seen in the history of both life and culture. These are just two of many theorists who, along with Stewart, are making the case for a much more interesting vision of evolution. It is a vision that has strong roots in science—all three of these theorists hew closely to conventional scientific mechanisms in their theories—but it touches upon all kinds of cultural and even spiritual concerns. Directionality is making a comeback, it seems, and progress is never far behind. For hard-core evolutionary biologists who brook no talk of such matters when it comes to their chosen profession, separating teleology from evolution must be a little like trying to play Whac-A-Mole at the local amusement park. No matter how many times it is beaten back by the powerful hammers of conventionality, it just keeps rearing its head again and again. Perhaps the naysayers are just fighting on the wrong side of history. As Stewart made clear to me that day in Arizona, progress and evolution belong together.

EVOLUTION OF A DIRECTIONALIST

Stewart, like all baby boomers, came of age in the aftermath of the two world wars. His father dabbled in theosophy and other spiritual trends of the day, a "dreamer" who invested in his son all of his own unfulfilled hopes. He gave him books to read, but not the usual teenage fare. Stewart's young mind was filled with significant twentieth-century figures like philosopher Karl Popper, spiritual teacher George Gurdjieff, and also an individual who seems to inevitably surface as a formative influence in the lives of so many Evolutionaries—Pierre Teilhard de Chardin. Teilhard's assertion that evolution follows a clear trajectory toward higher and higher levels of unity and organization planted an important seed, which made intuitive sense to Stewart even at that young age. "It seemed obvious," he told me. "It's the story of atom to molecules to replicating molecular processes to simple cells to complex cells and eventually to organisms."

In an alternate life, Stewart might have gone on to be an accomplished scientist. He certainly had the aptitude for it—as a teenager he excelled in both math and science. But his real talent was not science per se but what he calls "modeling," or discovering the underlying processes that make up any system. He discovered this skill at school, using a "classic mental-rational approach" to uncover the problem-solving techniques used by his classmates who did well on tests, and then intentionally working to enhance those skills in himself.

Later in his life, Stewart would use this same passion for modeling to theorize about the patterns and processes underlying not physics exams but evolutionary dynamics, in both nature and culture. But for the time being his teenage thought processes were turning toward another early talent—fishing. At age seventeen, he discovered the Great Barrier Reef, a fisherman's paradise, and he quit his studies and headed north to the Australian coastal city of Cairns. A fishing boat might not seem like the ideal training ground for a budding intellectual, but Stewart still had that scientific seed planted deep within. When the southeast trade winds blew and the fishing came to a halt, he would pass the idle days not at a pub or nightclub but at a local library. In the midst of those seemingly endless stacks of books, he was faced with a conundrum. How does one find the right books to read? True to his rational mentality, Stewart's answer was simple. He started at the number 0 in the Dewey decimal system, and began reading.

As fate would have it, philosophy comes quite early in the Dewey decimal system. And so this young Evolutionary-to-be found himself exploring the world's many systems of knowledge, enthralled by what he found. Around that time, he remembers reading an early book examining human culture in light of evolution, one that presaged later trends in evolutionary psychology. Despite its rather simplistic approach, the book opened Stewart's eyes "to the fact that evolution had the potential to provide a comprehensive framework that could essentially explain all the aspects of humanity."

That fundamental intuition drove him forward, and he read and read as the winds blew and blew outside, and when the winds died down, he would board his boat and chase Spanish mackerel up and down the Great Barrier Reef, a hundred miles out to sea. Perhaps it was the constant unpredictability of the elements, or his growing fascination with what was happening inside the Cairns library, or maybe it was simply that the romantic ideal of a fisherman's life began to lose some of its luster. But one day, sitting among the book stacks, Stewart realized that he didn't actually want to be a fisherman. He wanted to be a philosopher.

While Stewart never became a philosopher in the strict pro-fessional sense, he is very much a philosopher of evolution in a larger sense, someone who is taking data from science and trying to rethink how we interpret that data, philosophically and theo-retically. He has steered clear of traditional academic pursuits but has recently become affiliated with the Evolution, Complexity, and Cognition Group (ECCO), a transdisciplinary research in-stitute at the Free University of Brussels.*

As Stewart's intellectual life developed through his twenties and thirties, so did his career, eventually landing him in the Austra-lian government, where he fortuitously worked his way back to his early passion—fishing. Only this time he wasn't in the boat, he was at a desk, as manager of the Australian deepwater fisheries. In an era of rapidly declining fish stocks, it was an important, if difficult and thankless, job, but one that Stewart had a particular aptitude for. Indeed, it was during this time that his evolutionary thinking un-derwent a critical shift. As he observed the decline of the fish stocks and the struggles of the government to adequately respond, he was able to employ his natural capacity for modeling—diagnosing the underlying systemic behavior that was informing the dynamics he was observing in this consequential Australian industry.

How do you get fishermen to cooperate? If they work together and don't overfish, everyone wins, and the fish levels stay healthy and sustainable over the long term. But if one person overfishes, he or she spoils the party for everyone else. This is what scientists call the "free-rider problem." These fishermen become free riders, enjoying the benefit of everyone else's goodwill without incurring the same cost.

---

* ECCO was a spinoff of the Center Leo Apostel (see chapter 2), the research institute fo-cused on the evolution and emergence of cultural worldviews. This Brussels university has become something of a center for new thinking in evolutionary dynamics, systems theory, and the new complexity sciences, a tradition that goes all the way back to one of its more famous scholars, Nobel Prize winner Ilya Prigogine, who showed how certain systems or what he called "dissipative structures" can exhibit a tendency to spontaneously self-organize to higher levels of organization.

The same thing happens in evolution. Competition for scarce resources is critical to evolution's advance, but as we discussed in the previous chapter, nature doesn't favor just heroic individuals. Like Sahtouris, Margulis, and others, Stewart points out that other key trend in evolution—the increasing capacity of the living process to cooperate. This evolutionary trend is unmistakably present in the history of human societies as well. If we step back far enough from the details to see it, Stewart explains, the trajectory is clear: family groups cooperate to form small bands, which team up to form tribes, which coalesce to form agricultural communities, and on and on until we get to the extraordinary complexity of human societies today. Cooperation, cooperation, cooperation—all the way up and all the way down. It's the key ingredient that greases the wheels of evolution's advance. As he reflected on the powerful role of cooperation in the evolutionary process, Stewart believed he was beginning to understand the science behind Teilhard's description of the evolutionary trajectory. Cooperators naturally outcompete "go-it-aloners" in most systems, thereby providing a powerful evolutionary impetus for higher and more complex forms of organization to emerge.

So how *do* you get fishermen to cooperate? Certainly not by explaining evolutionary theory to them. No, they have to believe it is in their own self-interest to work together. In fact, for Stewart, the same fundamental principle is involved whether we are talking about bats, bees, bacteria, or basketball players. As he writes, "Cooperation emerges only when evolution discovers a form of organization in which it pays to cooperate."

But there was one more problem. In the case of bacteria, that process may take millions of years of trial and error before a form of organization is eventually discovered that rewards reciprocity and win-win arrangements. That's all fine if we're on biological time and have eons to burn. But that's not the case with human culture, when the seas are fast emptying, the biosphere is imperiled, and political conflicts threaten to erupt into nuclear warfare. What will enable a higher, more evolved form of organization to emerge when "free riders" are overindulging to the detriment of us all? How can we help further the cooperative process so vital to growth and development in any system?

It was this compelling combination of real world conundrums, evolutionary knowledge, and burning theoretical questions that was tumbling around in Stewart's mind one day in the early '90s at his home in Canberra. Suddenly things began to come into focus:

> These different issues were coming together as my mind started to move across levels. How do you organize a bureaucracy effectively so that people cooperate? How do you organize simple cells so that they cooperate together? How do you organize complex cells so they cooperate together? How do you organize multicellular organisms so they cooperate together?

Like new crystalline structures forming deep in his consciousness, a realization took hold, and all the problems and confusions and complexities began to resolve themselves. What was the key to this newfound clarity? In a word: *governance*. In the case of the nucleated cell (or eukaryote), for example, he realized that the reason it was so successful, from an evolutionary standpoint, was because its structure allowed for the emergence of a new governance function in the form of DNA, which managed the cooperative dynamics within the cell walls. It was this innovation that changed the playing field and made cooperators king in the palace intrigue of cellular politics. In the case of something as complex as international fisheries, the same function is needed, but at a global scale. Stewart explains:

> A general theory of organization started to emerge. It wasn't a logical progression; it all happened in a flash. I saw that the role of DNA in a cell was the same as the role of global governance

yet to come for the planet as a whole. I had these models to show why it was very difficult for cooperation to emerge and why self-interest generally trumped/precluded cooperation. But I realized that if you have a powerful entity that can control the activities of the smaller interacting entities, this powerful entity can appropriate some of the benefits of cooperation and feed them back to those who have contributed to that cooperation, thereby ensuring that they don't get outcompeted.

In this epiphany was the missing piece in his understanding of what makes cooperative evolution a success: governance and hierarchy. Stewart began to recognize the vital role of hierarchical governance as a spur to evolutionary advance. It allows us to do quickly with good policy what evolution once did slowly by trial and error—coordinate self-interest with the interest of the whole. This insight answered his biggest question: *Why* does evolution so clearly follow the Teilhardian trajectory? Cooperation organized by effective governance systems provides unmatched advantage over non-cooperators, and encourages the rise of higher forms of organization, all the way from the structure of cells to tribal councils to contemporary megacities to nascent planetary governance and beyond.

We may indeed live in a sociable cosmos, as we learned in the previous chapter, but it is also a cosmos that is going somewhere. And one way to track the trajectory of evolution is to track the trajectory of organized cooperation among evolution's constituent parts—in this case, us. The idea is exciting, but the practical implications even more so. In a world where so many major issues—global warming, drugs, corporate malfeasance, economic interdependence, financial regulation, nuclear nonproliferation, terrorism—involve some sort of failure to cooperate among constituent nations, the dramatic need for effective new forms of governance goes way beyond fisheries. The challenge of effective governance is not just another hot-button political issue. The need to help organize, incentivize, and otherwise oversee the many national and transnational entities and processes that now exist on this rapidly complexifying pale blue dot out on the spiral arm of the Milky Way is in fact an evolutionary imperative, a developmental challenge for the species itself.

Indeed, recognizing the trajectory of evolution is much more than an exciting epiphany. It's a tremendous responsibility. To begin to see the outlines of the path ahead means that we are responsible for walking down that path. We are beginning to take on our own shoulders the burden of our future, and to appreciate that a richer understanding of evolution provides real clues as to the direction of human culture—not as some straightjacketed, predetermined pathway but as a compelling trend of history, providing essential context for our global challenges that we would not do well to ignore. To see the directional nature of evolution is to make a critical link between the deep past and the possible future, and to connect the cultural drama of our human concerns to the far more vast drama of biological and ultimately cosmological progress. It is to place the choices we make at this moment in history in that spectrum of critical transitions that have punctuated the evolutionary trajectory for billions of years and made it such a dramatic, remarkable, unpredictable affair.

In "The Evolutionary Manifesto: Our Role in the Future Evolution of Life," a paper he published in 2008, Stewart calls on "intentional evolutionaries" to "recognize that they have a critical role to play in driving the evolutionary transition and the future evolution of life," and to "use the trajectory of evolution to identify what they need to do to advance evolution." In part, this means that we have to find ways to facilitate global organization and cooperation on levels never tried before. We have to find a way to coordinate our self-interest—individually, tribally, nationally, and so on—with the interest of the whole species and the living processes of the planet. What this typically points to, in living systems, he explains, includes "the near eradication of activities such as the inappropriate monopolization of resources by some members, the production of waste products that injure other members, and the withholding from others of the resources they need to realize their potential to contribute to the organization."

The forces of pure competition cannot negotiate such delicate evolutionary challenges. But neither can cooperation for its own sake. Cooperation always needs a reason, a compelling and clear context and an organizing hierarchy—an overarching structure that gives meaning, purpose, and direction to the cooperative impulse. In our current global situation, some of that may come from the need to survive as a species, to face up to the genuine transnational threats to our civilization, be it climate change or some other clear and present danger. But an evolutionary context is not just about problems; it's about possibilities. And if we start to see in the directionality of evolution a potential to be embraced rather than just a problem to be solved, then we are beginning to cast off a blinkered view of history and embrace a larger motive. If we understand the reality of the trajectory of life, as Stewart suggests, we can project it forward and see the potential of the human species to eventually participate in cooperative arrangements that may extend far beyond even our own planet and species.

But before we start drawing up charters for some galactic UN, it's good to keep perspective. Even if we were to agree on a clear directionality in evolution, there are no guarantees in the process. There are no historical inevitabilities, at least not in sophisticated evolutionary circles, and no prizes for mere participation. The fact that we can begin to recognize the broad trajectory of evolution does not mean that enough people will actualize its potential by walking down that path.

What makes theorists like Stewart particularly interesting is that he is drawing his conclusions about directionality and progress in evolution primarily from science. Sure, one can venture outside those parameters into more philosophical, spiritual, theological, and metaphysical explorations, and sometimes there is good reason to do so. But evidence for directionality is not hard to see, even within a close and careful reading of nature's scientifically revealed patterns. In this respect, Stewart reminded me of another self-deprecating, hardheaded rationalist back on this side of the Pacific Ocean— Robert Wright, the American Evolutionary who has been shaking up the establishment with his firm conviction that there is nothing random about the way in which not only nature but also culture is progressing.

SUGGESTIONS OF PURPOSE

In 1500 BC there were around 600,000 autonomous polities on the planet: 600,000 separate groups of people who had their own independent form of government. Care to take a guess at how many there are today? The answer, according to evolutionary theorist Robert Wright, author of *Nonzero: The Logic of Human Destiny*, is: fewer than 200. That's quite a change in the last 3,500 years. Now, here's the important question: Given this fact, would you say that culture over the last three millennia is progressing or regressing? Of course, it's impossible to answer the question based only on this one simple fact. But barring an apocalyptic scenario, I'd say that most people would choose door number one—culture is progressing.

I first met Wright at the 2004 Parliament of the World's Religions in Barcelona, Spain, a unique gathering that brings together leaders and laypeople from practically every religion on the planet once every five years to engage in dialogue and compare notes. The setting for our meeting was itself a powerful example of Wright's field of specialization: cultural evolution. If one thinks back for a moment to all of the religious violence of the last several thousand years and then imagines the scene in Barcelona—nine thousand monks, priests, teachers, bishops, imams, theologians, scholars, saints, Sikhs, seekers, sheiks, priests, pagans, and laypeople from seemingly every religious tradition on the planet, all cooperating and getting along exceedingly well, this alone might be enough to convince one that some kind of cultural progress is occurring in the currents of history.

Wright, who was also in part inspired by the writings of Teilhard de Chardin, claimed that evolution had an unmistakable direction, and that its trajectory could be clearly seen in the development of both biology and culture. He based his ideas on a term he borrowed from game theory, an obscure branch of mathematical studies that John von Neumann helped make famous in the 1940s. A "non-zero-sum" interaction is one in which both parties benefit: a win-win arrangement, as opposed to a "zero-sum" interaction, or a win-lose arrangement. If I produce a widget and sell it to a happy customer, we have engaged in a non-zero-sum interaction. I received money and the customer received a valuable widget. Both benefit; both are happy. It's win-win. And Wright's application of this idea to evolution was simple but brilliant. He suggested you could essentially track the evolution of culture by tracking the increase in "non-zero-sumness" throughout human history. Thus, as culture develops, human beings interact, economically and socially, in ways that produce more and more win-win arrangements for more people in larger networks in increasingly complex ways. Technology is a primary driver in this process, as it allows these links and connections to be established across wider and wider geographical areas. This increasing interdependence ties us all together in ever more vast webs of non-zero-sum relationships until . . . well, no one knows exactly where it's all headed, but the direction is clear. Our own self-interest is increasingly connected to the well-being of civilizations and peoples half a world away. It's not exactly the Age of

Aquarius reloaded, but as a clear direction for cultural evolution it's quite definable and defensible. As non-zero-sumness grows throughout history, a sort of directional moral trajectory can be seen in the evolution of culture. We are less likely to kill and make war on those with whom we are engaged in win-win relationships. Planetary goodwill is slowly prevailing over individual and group hatred.

Wright's work calls to mind columnist Thomas Friedman's observation that it is rare for two countries with Golden Arches to go to war with each other. Of course, that doesn't have anything to do with any moral halo bestowed upon local populations by the presence of Big Macs; it is merely a metaphor showing how economic interdependence tends to lead to cooperative behavior rather than bloodthirsty killing. It becomes part of the economic self-interest of a nation to maintain peace. And while this self-interest may not have the same spiritual significance as selfless altruism, Wright argues that self-interest arising from non-zero-sum interactions is a powerful evolutionary driver in culture. The result is a snowballing globalization in which the fates of people around the world are increasingly linked—first in small clans and tribes, then in larger and larger collectives, and eventually in global non-zero win-win relationships. And he suggests that this direction to human evolution may be leading to a moral culmination, inciting us to develop the "infrastructure for a planetary first: enduring global concord."

> Historically, the amity, or goodwill, within the group has often depended on enmity, or hatred, between groups. But when you get to the global level, that won't work—assuming we don't get invaded by Martians or something. That cannot be the dynamic that holds the planet together. So we're facing a new kind of challenge. . . . In the past, we moved from the hunter/gatherer village to the agrarian chiefs to the ancient state, so there are

precedents for the expansion of solidarity between people. But what would be unprecedented is to have this kind of solidarity and moral cohesion at a global level that did not depend on the hatred of other groups of people. That would be a singular accomplishment in the history of the species.

Wright argues that this non-zero dynamic was also at work in the evolution of organic life, which echoes many of Stewart's points about cooperation being critical to the evolution of biological systems. But Wright's focus is on social evolution. His vision is optimistic, but it is a grounded optimism, one that offers no inevitabilities, no idealistic promises about the outcome of the human experiment. However, recognizing directionality in evolution does inevitably lead one down interesting avenues of conjecture. For example, when asked if the directionality of evolution implies a larger purpose to the process, Wright concedes that it is "at least suggestive of purpose. . . . A scientific worldview gives you more evidence of some larger purpose at work than most scientists concede." Wright even wrote that "if directionality is built into life . . . then this movement legitimately invites speculation about what did the building."

That's about as far down the rabbit hole as Wright will venture when it comes to spiritual matters. Perhaps it is because of his experience growing up in a Southern Baptist family of a very conservative persuasion. "My parents were creationists," Wright (a fellow Oklahoman), recently told the *New York Times Magazine*. "They brought a Baptist minister over to the house to try to convince me that evolution hadn't happened." Obviously, this attempt was unsuccessful. Wright's evolutionary vision is strongly rooted in science. But he is trying to carve out room for legitimate speculations about morality, purpose, and even spirit in the framework of a rational, naturalistic picture of the cosmos. We don't need to resort to supernatural explanations to find evidence that there is higher meaning and purpose to human life. We can see it right here in our own backyard, displayed prominently in the extraordinary trajectory of the evolutionary process.

If there is a wall between science and religion, then Wright's work is like a large gap in that wall, excavated by him. He is standing atop the wall, inviting passersby to temporarily step through the gap and embrace another perspective. But he himself remains perched up there, refusing to come down on one side or the other. Like Stewart, he is a rationalist through and through. Perhaps partly as a reaction to his southern Baptist upbringing, he is highly suspicious of traditional mythical belief systems. Yet he also feels that a purposeless, meaningless cosmos does not come close to describing the processes and results he sees at work in evolution. He is a staunch Darwinian—a great believer in natural selection as the mechanism of evolution. It's just that, from his point of view, that process moves evolution in a progressive fashion toward intelligence and moral development. In fact, he sees in it a directionality so remarkable, so compelling, and so clear that it suggests that there is a larger purpose at work in the unique logic of this historical unfolding. It is a nuanced position that has won him mainstream acclaim, but little love from evolutionary scientists, who have regularly accused him of opening the door to dangerous metaphysical speculations, raising fears that the forces of superstition are about to crash the gates of rationality.

In 2009, Wright's former employer *The New Republic* asked popular evolutionary biologist Jerry Coyne to review Wright's latest book, *The Evolution of God*. The title of the review said it all: "Creationism for Liberals." Coyne pulled no punches in his condemnation of Wright's fence-sitting. To be called a creationist, even a liberal one, is of course, one of the worst accusations one can throw at an evolutionary theorist. Coyne's choice of words captures just how unwilling or unable some are to distinguish between those philosophies of evolutionary purpose, meaning, directionality, and even spirituality that unambiguously represent a pro-evolution point of view and those that move us backward toward premodern worldviews and seek to undermine the integrity of evolutionary science altogether. In fact, it perfectly illustrates the confusion and conflation of these two vastly different approaches. Change is afoot, new philosophies of evolutionary directionality are coming soon to a bookstore near you, but the fears and failures of the nineteenth and twentieth century still, understandably, loom large in the collective mind of the twenty-first. Evolution, as always, takes time.

THE DARK AGES, THE GARDEN OF EDEN, AND THE TWENTIETH CENTURY

"Evolution meanders more than it progresses," observed Michael Murphy, pioneer of the Human Potential movement. Wherever evolution is going, at first glance it seems to be taking its own sweet time, and getting quite distracted along the way. If self-reflective hominids were really where evolution was headed, couldn't we have moved the process along a bit faster? Do we need 350,000 species of beetles? Biologist J. S. Haldane is famous for saying that if biology taught him anything about the mind of the Creator, it was that he or she had an inordinate fondness for beetles.

When it comes to cultural evolution, the same principle applies. If we follow too closely the rise and fall of civilizations and the turning tides of history from one century to another, it's not always easy to detect real development—moral or otherwise—in human nature or in institutions of human culture. It's only when we step back from the details and look at the larger trajectory of human civilization that progress becomes much more obvious and discernible. But even then, progress is unpredictable: different cultures develop at different rates and move forward at different times. And then there are these black spots on the historical record that always get raised as refutations of the whole idea. "How could cultural evolution possibly be true, given that X happened?" goes the usual argument. Perhaps the most common witness for the prosecution in this case is the so-called Dark Ages. Isn't this period a black mark on the historical record that refutes cultural evolution?

In fact, it is not, for several reasons. First, it bears repeating that evolution meanders more than it progresses, and some of that meandering is sure to include major setbacks and all kinds of minor ones. Cultural evolution, in more sophisticated circles, is not imagined to be a simple upward slope of ever-ascending progress and development. Evolution simply doesn't work like that, at any level. Goethe once wrote that "progress has not followed a straight ascending line, but a spiral with rhythms of progress and retrogression, of evolution and dissolution." Crisis, catastrophe, and disaster are each a major part of evolution's rich repertoire. Remember what we learned in the previous chapter—stress creates evolution. In fact, it is often life's epic setbacks that set the stage for a new evolutionary advance. So it was with the extinction of the dinosaurs, when mammals were finally free to have their day in the sun. So it has been with almost every one of the major extinctions the earth has faced. That is why "evolutionary evangelist" Michael Dowd, whom we will get to know later in this book, likes to say "Love the bad news." Somewhere, amid the crisis of the moment, the stage is being set for great leaps forward. In fact, that basic pattern can be seen in the case of the Dark Ages as well. Wright, for example, argues that this particular era was not quite as dark as generally imagined. He suggests that the massive decentralization that occurred, along with the economic and political experimentation of the time, set the stage for the great leaps forward in European culture in the later centuries. That doesn't mean that we should embrace failure or pretend that the tremendous suffering endured as a result of the fall of the Roman Empire is somehow a good thing. Far from it. But the simplistic conclusion that these difficult times in history represent a wholesale refutation of the larger trend of cultural evolution is not necessarily supported by the evidence. When I asked Wright about the meandering nature of cultural evolution, he agreed:

It's kind of like biological evolution. There are different directions that different lineages take. There is a lot of contingency, in that it was not at all clear a billion years ago which lineage would be the one that led to higher intelligence. On the other hand, if you just check back every hundred million years, the tendency was that at that point in time the most intelligent form of life was more intelligent than the most intelligent form of life had been a hundred million years earlier. And there were such epic setbacks that even that might not be true for a couple of hundred million years. And yet still the larger trends ultimately prevailed until you have intelligent, reflective, self-conscious, morally aware life. I think that property of higher intelligence was in the cards, so to speak—even though there were going to be all of these setbacks. I think moral progress in culture is similar.

A second example commonly used by critics of cultural evolution is closer to home, historically—the world wars of the twentieth century. Wasn't the twentieth century the most violent in history? Wasn't the Holocaust a perfect illustration of the naïveté of believing in cultural progress? Wasn't Germany supposed to be the most advanced culture in Europe? How can we pretend that culture is moving forward amid the nightmares of the twentieth century?

This is perhaps a more difficult point to argue, if only because the pain and trauma of the catastrophes are so near. These disasters are still part of the living cultural memory of our time, only one or two generations removed, and so their proximity tends to fog our historical perspective. I can't tell you how many times I've been rudely upbraided for even suggesting that the twentieth century was not necessarily the most violent century in history and that we are actually becoming more peaceful as history progresses, despite clear and painful setbacks. But here again, I would suggest that the evidence is on the side of the directionalists.

Let's look at the most important barometer of cultural evolution: moral progress. What about moral evolution? Are we becoming more peaceful over time? More caring and empathetic? More able to extend our concern to others, to appreciate the perspective of others, to walk in their shoes, so to speak? This is difficult to measure. We can argue for evolution in economic circumstances, political institutions, technological capacity, and societal complexity, but when it comes to assessing cultural evolution, those material measurements need to be married with a less material quality—the evolution of our inner lives. For the most part, I will address this dimension of evolution in later chapters, but for our purposes here we can still ascertain clear evidence for moral progress, even amid the carnage of the twentieth century.

The convulsions of the twentieth century, like those that began the Dark Ages, were the result of one kind of political organization giving way to another in the march of history. In the first millennia, it was of course the fall of the Roman Empire, and in the twentieth century it was the transition from the colonial empires of Europe to the new forms of political order that would emerge in the latter half of the century. Indeed, we can see how both world wars gave rise to significant new attempts at global governance—first the League of Nations, and then the United Nations, NATO, and the so-called Bretton Woods economic arrangements. Peace tends to follow in the footsteps of economic and political organization. As Stewart's work makes clear, we cannot really hope to have global peace without first hoping for some kind of decently functioning global political and economic institutions. And following Wright's non-zero logic, the more we are engaged in win-win relationships with others, the more we are likely to see ourselves as being "in the same boat" and to extend our circle of care and concern—to see our own self-interest as connected to and coordinated with the self-interest of the larger community. In this sense, we can ascertain a certain level of moral progress in history simply in the fact that these "circles of concern" have extended from clans to tribes to city-states to nations, and today, many people consider themselves to be world citizens first and foremost. Our perspective has evolved, from *egocentric* to *ethnocentric* to *nationcentric* to *worldcentric*. Never before in history have so many people considered themselves to be simply members of the human species or citizens of the planet as their primary self-identity. Such an idea would have been unthinkable a thousand years ago. And as political theorist Thomas Barnett recently reminded me, it's the first time in history when all the major European powers are relatively peaceful, prosperous, and moving toward integration, even in the midst of current economic challenges. There is little concern about the possibility of another major war between great powers. That's a huge step forward and a recent phenomenon. We can say the same thing about Asia. India, Japan, and China are all relatively prosperous, peaceful, and developing. This has never happened before in history.

None of this denies the tremendous challenges we face and the many contingencies that make progress tenuous. And it's not to ignore some of the unfortunate and inevitable by-products of increasing levels of development. They bring their own complex challenges that would have been unthinkable in previous eras: terrorism and climate change being two that fill current headlines. But as we move further into the twenty-first century and the memories of the early twentieth century fade, I suspect we will be able to more accurately place the horrific events of those times in the context of larger patterns of history. Whatever the case, they stand as living reminders that any cultural progress is contingent upon many factors and that man's inhumanity toward man can never be underestimated.

One scholar who has done much to help us place the violence of our own time in context is Harvard archaeologist Steven LeBlanc. Which brings us to a third argument often used to refute a directional view of cultural evolution. We might call it the Garden of Eden myth—the conviction that there was a peaceful, idyllic period somewhere in the mists of our prehistory. This notion is especially pernicious in some progressive circles. Weren't we more peace-loving, happy, and free somewhere back there in the past, at some point before the evils of the modern world corrupted the hearts of men and women? Aren't humans deeply good? Haven't we just been corrupted by all of these layers of socialization and civilization? Weren't we once in tune with the natural world? We don't need evolution, such arguments go, as much as we need to undo the harm the modern world has inflicted upon the inherent goodness of human nature. We don't need to evolve forward in history; we need to undo history!

LeBlanc's scholarship, captured in his book *Constant Battles: Why We Fight*, is one of the more devastating objections to this idea. "Prehistoric warfare was common and deadly," he writes, "and no time span or geographical region seems to have been immune." More recently, Steven Pinker's *The Better Angels of Our Nature* has made a thorough and convincing case for the decrease of violence over the course of human history.

There is nothing new about the Garden of Eden myth. Since the moment "civilization" first became a word used to describe human culture, people have been longing to return to a time before whatever the present time period happened to be. We have always looked back to an era before the current toil and strife and imagined a more bountiful and peaceful Golden Age. Whether it was Rousseau in the seventeenth century claiming that without civilization, man would not know hatred or prejudice, or the recent movie *Avatar* that celebrated the peaceful, ecologically enlightened, spiritually rich, technologically poor forest-dwelling indigenous population of an alien planet, the idea of returning to a more peaceful, happy, socially free existence is a powerful and enduring cultural myth.

"The world is too much with us," lamented Wordsworth in his famous sonnet, exclaiming that he would rather be "a Pagan suckled in a creed outworn." For anyone who has spent a good deal of time lamenting the devastating effects of industrialization upon the beauty of the natural world, the poet's words have a lingering power. Indeed, the rose-tinted sentimentality for the past has always had a strong currency in our culture. And today, with the sheer velocity of technological change, the dislocations caused by rapid globalization, and the many ecological challenges we face as modernity's promise is embraced by billions of new people around the world, it makes sense that a certain longing for simpler times and a "creed outworn" would pervade the cultural zeitgeist.

Celebrated author Riane Eisler has been one of the most effective proponents of the Garden of Eden hypothesis in our own time with her 1987 book *The Chalice and the Blade*, which presented scholarly evidence for a prehistorical civilization in which matriarchy was the power structure, Goddess worship was the spiritual practice, and humans lived in a peaceful, harmonious, and nurturing way. Women were empowered but men were not subjugated and war was relatively unknown. She used as examples of her thesis the prehistoric cultures of Ireland and the early civilizations of ancient Crete. Eventually, according to Eisler, the rise of male-controlled "dominator" societies brought an end to this ancient cultural experiment, and violence returned to human affairs. Eisler is an accomplished author and her book caused quite a sensation, prompting many to reconsider the history we learned in school about the early civilizations and our prehistoric past. Over time, Eisler's assertions about our matriarchal heritage have come under strong criticism, with many claiming that her scholarship is highly interpretive and that the archeological evidence ambiguous at best. In most cases, it appears that war and violence were a common part of all prehistoric cultures.

I suspect the historical evidence will ultimately tell us more about the brutality of earlier civilizations than their idyllic impulses. But whatever the merits of Eisler's conclusions, there is a very clear motive behind her work. In fact, she has been quite transparent about her intentions. She is trying to show that violence and domination of one race by another, or one gender by another, is not an inevitable part of human affairs. Her work has been an effort, she writes, at "dispelling the notion that war is natural." Raised in the shadow of the Holocaust, Eisler is attempting to help humanity move beyond the "dominator" ethos that still conditions so much of our culture today.

From a nonevolutionary perspective, I might understand her concern. But in the context of an evolutionary worldview, which allows us to chart the evolution of our own social identity from egocentric to ethnocentric to nationcentric to worldcentric and beyond, we have no need to reach backward to find a utopian model for a better future. Indeed, we can find all the inspiration and faith we need for the future not in any particular time period or event in the past but in the overall trend of history itself. That is what gives us the energy to move forward. It does not mean that missteps haven't been made or that important breakthroughs at one stage of cultural evolution don't provide the seeds for problems at the next. Every step forward presents its own unique challenges. Evolution is not a simplistic, all-or-nothing affair. It meanders more than it progresses, but it *does* progress. And we need only look back to get a sense of that progress and direction. "Because we can look back and see the pattern," evolutionary theorist Beatrice Bruteau writes, "we can legitimately extrapolate and project the pattern into the future." There is nothing mystical about this. Whether it is John Stewart's revelation of cooperation and governance as providing the spur of evolution's trajectory or Robert Wright's examination of the unfolding social logic of cultural history, we can find a powerful, evidence-based argument for the reality of directionality in evolution. I have only begun to explore the implications of such a finding, but they are significant. They start first and foremost with the story we tell ourselves about our past and our future. No longer aimlessly adrift on a cosmic sea, we awake to find ourselves drifting forward. We can draw on that evolutionary trajectory not as unsupported confidence in the certainty of future advance but as a deep, rational conviction that amid the many headwinds of the present lies a subtle but unmistakable current, a direction, a positive indicator of a future trajectory. If we can see it, we have to help make it so.

CHAPTER SIX

Novelty: The God Problem

Human consciousness is not merely an emergent phenomenon; it epitomizes the logic of emergence in its very form. . . . Consciousness emerges as an incessant creation of something from nothing, a process continually transcending itself. To be human is to know what it feels like to be evolution happening.

—Terrence Deacon, "The Hierarchic Logic of Emergence"

Why are there beings at all, instead of Nothing?" the philosopher Martin Heidegger famously asked. This question, he proposed, is the root of all philosophy, and it is a question that still confounds philosophers, scientists, and Evolutionaries of all stripes. We have explanations for how one thing transforms into another over the slow march of time. But evolution's greatest mystery still remains: How does something come from nothing? And ultimately, it's not just a question about the origin of things but about novelty of all kinds. How does anything *new* get created? How does something entirely novel come into existence? In this world of change and flux, what is the source of unexpected creativity?

We have studied creativity in the human species for years, and yet still it remains elusive, and when we get to the natural world, the question looms even larger: How do you explain the appearance of novelty in this awe-inspiring universe around us? The religious traditions had it easy. Novelty was simple—it came from God. Take the Genesis story, in which God created the heavens and the Earth, just as they are today. He was the creator; not humanity, not nature. God did it all together, all at once—no changes, no additions, nothing that would implicate nature or humanity in a creative act. Similar narratives are found in the Hindu Rig Veda, the Quran, ancient Egyptian mythology, and the sacred texts of many other cultures. It was a neat and uncomplicated picture; unfortunately, it just happens to have little to do with the world as we understand it today. Today "Mother Nature" has taken over the role once reserved for deities. We live in a time in which it seems that nature is revealed to be increasingly creative with every scientific advance, every new discovery, each breakthrough insight into the mysteries of physics and biology. And humanity itself has become a race of creators extraordinaire, giving birth to all manner of novelties, from Shakespeare's sonnets and Rembrandt's portraits to iPhone apps and virtual worlds. In this sense, we have truly become, as *Whole Earth Catalog* founder Stewart Brand said, "as Gods." But for all our progress, we are still only studious apprentices of nature's masterwork. We have neither been able to fashion life from seeming lifelessness nor wring self-reflective consciousness out of the interactions of matter. Perhaps one day these mysteries will be within our grasp, but for now, we cannot match evolution's inherent creative power.

From a certain perspective, novelty itself is something of a novelty. Of course, now we know there have been new emergences ever since the first great emergence created our cosmos out of nothing, and that human culture has been defined by a series of unexpected creative leaps. But for the most part, these leaps were so widely interspersed that we barely noticed the phenomenon within our short lifetimes. Indeed, it is worth noting that the issue of human creativity as a subject worthy of study wasn't even raised in the culture at large until the nineteenth century. Before then, it was not studied by philosophers, pondered over by artists, or simplified into steps by self-help authors and organizational consultants. Sure, there were occasional writings about the "muse" of the artist or writer and individuals have obviously been extremely creative since the beginning of human existence. But creativity as a subject of inquiry didn't really exist until the nineteenth century. So what changed? The answer is that as the industrial revolution starkly transformed the Western world, human beings were confronted with change and progress in history on a much smaller time scale. We saw the possibility of improving our lives and creatively working to enhance society. Human progress suddenly seemed easily discernible in the space of one lifetime. Of course, downsides emerged as well. Factories were dark, dirty, and unsafe. Masses moved from the rural existence that previously marked human life to the chaos of overcrowded cities and slums. But amid the challenges, change was speeding up. Like never before in human memory, the life of the son was almost certain to be quite different than the life of the father. Cultural development was making its own mechanisms transparent to us.

In the sciences, too, a new world was being unveiled. Nature was yielding secrets of evolutionary change, not to mention a world of natural laws and a whole new universe of subtle but discernible processes working quietly in the depths of life and matter. Our understanding of time itself was going through a revolutionary change as we began to move out of a cyclical, seasonal, more agriculturally oriented conception into the industrial, modern sense of moving forward in history, as a nation, and as a species. Change, process, progress, development, evolution—all of these words came into vogue in the nineteenth century and all of them exist in the same intellectual ecosystem as the idea of creativity.

When it comes to evolutionary theory, few subjects play a more central role than creativity. What gave rise to new forms, new species, new functions and capacities? How did life emerge? How did self-reflective human consciousness emerge? These are the questions that confront the evolutionist, and every theory or approach to understanding the evolutionary process, from Darwin to Dawkins, is, in some sense, an attempt to explain the source of this miracle called novelty. Novelty is like the Rosetta stone of evolution. If you can explain it, even partially, then you have achieved something remarkable, and perhaps brought humanity a little closer to unlocking one of the core mysteries of life and existence. And in order to more fully realize a worldview organic to an evolutionary age, we need to pay attention to the underappreciated mystery of novelty—both in the depths of ourselves and in the depths of the cosmos.

In this chapter I want to explore the nature of novelty through the work of some of today's most interesting evolutionary theorists. For the most part, I will focus on the scientific inquiry into the issue, looking at this fascinating subject through the lens of several of today's leading thinkers in complexity theory. Because the subject of creativity also calls to mind the objections of those in the intelligent design camp who claim that current biological theory is insufficient to explain novelty in the emergence of life, I will unpack their concerns. But to help frame the overall theme of this chapter, I will first introduce one of the more provocative and interesting voices in the world of science—a man who had thought a great deal about if and how a godless universe has the power to create.

THE CAFÉ AT THE BEGINNING OF THE UNIVERSE

Certain people just have a gift for capturing a particular essence of the natural world. For example, when I watched *Cosmos* as a young boy, I was captivated by Carl Sagan's masterful rendering of science's latest forays into astrophysics and cosmology. His unique talent lay in communicating both the latest knowledge of science but also what was happening at the very edges of the field, those areas that were still unexplored, unknown, and unseen by the probing lenses of our telescopes and microscopes. His gift, in other words, was not just transmitting the latest answers to the mysteries of the cosmos but bringing to light the critical questions that defined the frontiers of science, the permeable membranes between what we know and what is just beyond our collective grasp. It was a lesson in both the discoveries of science and the nature of the quest for knowledge itself. On that journey, I understood, the right question may be much more important than any given answer.

One person who possesses an analogous skill when it comes to drawing our attention to the novelty inherent in the cosmos is author Howard Bloom. Bloom is something of a novelty himself in the world of intellectual attainment. His books, *The Lucifer Principle*, *The Global Brain*, and *The Genius of the Beast*, are fascinating portraits of cultural, biological, and cosmological evolution. I was familiar with Bloom's prolific writings and had been struck by his synthetic and adventurous mind. With a deep knowledge of history and a broad expertise in many fields, he was able to move easily between different sciences and wasn't afraid to mix hard facts with cultural analysis, weaving disparate knowledge systems together in a way that left the astute reader enriched (if a little dizzy).

A lover of science as a child, Bloom started studying cosmology, theoretical physics, and microbiology at the age of ten. By the ripe old age of twelve, he had built his first Boolean algebra machine. At sixteen, he worked at the world's largest cancer research lab, the Roswell Park Cancer Institute in Buffalo, New York, and at twenty, he was researching B. F. Skinner's "programmed learning" at Rutgers. And then his career took an unexpected turn. Caught up in the cultural changes of the late '60s and '70s, Bloom temporarily left behind the world of science and became a rock-and-roll media merchant, active in the superstar PR universe, with friends like John Mellencamp, George Michael, and Michael Jackson. He spent seventeen years doing what he describes as "fieldwork in the mass passions, the mass behaviors, that make history," returning to his science-inspired roots only when the onset of a mysterious illness not unlike chronic fatigue syndrome afforded him time to begin writing. I had heard that Bloom was largely nocturnal and kept crazy hours, and it was true that e-mails from him were more likely to come at three o'clock in the morning than at any reasonable time in the light of day. In fact, my request for a meeting was met with an unusual response: "Any time between 8:00 p.m. and 1:30 a.m."

He met me at the door of his Brooklyn brownstone and shook my hand, a diminutive man with a sparkle in his eyes. He led me upstairs several floors and into his "office," which the layperson might understandably confuse with his bedroom. With shelves and bookcases filled to the brim, it looked like a decade's worth of random objects had been scattered around the room and simply left to mark time. As we settled in for our discussion, Bloom joked about "the legendary mess." Eccentric without question, he is also warm and engaging, and as we began to discuss evolution and creativity he didn't waste any time getting to the point.

"Now, I am a stone-cold atheist. *Period*. I am a stone-cold scientist. *Period*. But there is more to this universe than science so far has been willing to grapple with. In a sense, those who argue for things like intelligent design are calling our attention to a real problem with how we understand this universe. How does a godless universe create?"

As Bloom continued to speak, I realized that I was already facing a conversational challenge that had no doubt confronted many a visitor to the Bloom brownstone. In order to answer a question, my erudite host felt like he needed to provide context—usually about ten minutes' worth, not including a few diversions along the way. But what I began to piece together from Bloom's monologues was his deep interest in the issue of creativity—and his concern that many don't appreciate the extent to which radical novelty has been at the core of the universe's evolution right from the start. As a scientist, he had no sympathy for those who are using notions like intelligent design to try to squeeze a traditional God back into a natural universe. But that doesn't mean that there aren't significant areas where science has yet to fill in blanks as to how A became B and B became C in the grand journey of cosmic history.

"Imagine that you and I were sitting in a café at the beginning of the universe, at a coffee table at the edge of the nothing, or what one person calls the 'potentia,'" he said at one point in our conversation, initiating a lengthy metaphorical story. "You are full of imagination and I'm crusty and conservative. Suddenly, you make this nutty prediction: 'Look over there' (of course, there is no 'over there,' but this is a metaphor). 'That spot over there is suddenly going to erupt into something that is infinitesimally smaller than a pinprick and it's going to be a burst of time and space, an explosion of speed. It's going to come from nowhere and—' At that point I stop you, and say, 'Look, Carter. You and I have been sitting around the potentia for as long as there's been a potentia. There never has been time, and there never has been space, and there never has been speed. You're talking about something coming from nothing. Now, remember the first law of thermodynamics, the conservation of matter and energy. *Nothing* comes from nothing. Something only comes from something. That's a basic law of science.'

"And then, all of a sudden—*Whammo!*" Bloom gestures in the air above his bed to emphasize the point as he stares me down. "This infinitesimally tiny little thing comes into existence and it's got a whole universe implicit in it. I mean, there's never been time, space, or speed, and this is a *manifold* that comes rushing out at a speed that defies belief. Guess what? You turned out to be right."

Seated at our metaphorical coffee table, Bloom takes me through several more phases of the extraordinary creativity that was present right from the birth of the cosmos. "We could go through the whole history of the cosmos this way," he explains. "And at every step some astonishing, unbelievable thing is happening. And every one of those defies what we currently know about the laws of logic and science. The God problem is this." And with that Yoda-like statement, Bloom finishes his café metaphor with a rhetorical flourish.

The God Problem is also the name of Bloom's most recent book exploring the topic of creativity. Such language is likely to provoke a few frowns from evolutionary scientists who are desperately trying to keep the phraseology of religion out of their field. But Bloom's invocation is for effect only; I've rarely seen anyone who loved science so deeply or was so well versed in this history of the field. And yet, he can't help but play the provocateur when it comes to issues close to his heart—such as the inherent creativity of the cosmos.

Bloom is deeply attuned to the latest scientific picture of the universe and the remarkable degree of development and creativity that has unfolded in the depths of matter over the last 13.7 billion years. He points out that the more we know about the universe, the more we understand the creative leaps our cosmos has taken in its long journey from hydrogen to humans, the less random it all seems. In fact, he feels that there is a remarkable order to the process, some kind of open-ended, creative intelligence that is informing cosmic evolution, a flexible but influential metastructure to the chaotic unfolding. In a word, there seems to be a *design* to the universe that we have yet to fathom, but not the kind that comes from God. Bloom is talking about deeper layers of science—fundamental ordering principles that might account for the surprisingly nonrandom character of the cosmos, axioms embedded in the fabric of the nature whose existence we can see hints of when we examine past leaps of cosmic creativity.

This is not, he tells me, a "six monkeys at six typewriters" universe. That phrase comes from that old and overused piece of conventional wisdom that a group of monkeys typing randomly at keyboards would eventually, given an infinite amount of time, bang out the complete works of Shakespeare. This has been a popular metaphor, used often in the evolution debates, with many variations on the basic "dumb primates producing intelligent things accidentally" theme. The idea has been employed to show that seemingly random processes can lead to order and even intelligence, given vast amounts of time and the right circumstances. Unfortunately, the probability, experts tell us, of a collection of typing primates producing even one *Romeo and Juliet*, one *Macbeth*, or even just an episode of *Melrose Place*, is near zero. (Incidentally, a clever experimenter tried this in real time, setting up a room with monkeys and typewriters and leaving them alone for some time. The monkeys ended up using the typewriters as a toilet and the only work produced was a series of pages with the letter *s*.)

Drawing further on in his café metaphor, Bloom points out that even in the early days of the cosmos, a truly remarkable degree of coherence, order, and structure were informing evolution. In his book *The God Problem*, he describes the creation of the first elements in the periodic table, several hundred thousand years after the Big Bang:

> How many kinds of atoms does a cosmos of zillions of particles sliding into each other's arms produce? If things were really random, the species of newly born atoms should be wacky, crazed, and without end. But in this universe, wild, weird, wacky, and endlessly crazed is not the way things go. Not at all. . . . Look, even just two cubes tossed around in a cup, dice, have thirty-six possible outcomes. How can an entire cosmos seething with more protons, neutrons, and electrons than we have words to describe, how can a universe of nearly infinite dice and nearly infinite tosses, produce just three varieties of atoms? . . . This is staggering conformity and self-control. . . . It is not mere trial and error. . . . So what is it? It's the paradox of the supersized surprise. It's the mind-snarler at the core of cosmic creativity.

Creativity is the God problem. We have to come to terms, Bloom tells me, with these "material miracles" that are present at every stage of the evolution of the universe. For Bloom, science is best served when our sense of awe, wonder, and astonishment at the workings of nature is heightened. To drive home the point, he comes back to one of the essential elements of science—predictability. If we can't predict what the next great creative leap is going to be, he tells me, in this universe whose most distinct feature is its ability to manufacture one grand creative leap after another, then have we really understood this cosmos? Have we explained the natural world, understood its workings, grasped its most salient and essential features?

He points out that the problem of creativity is actually a relatively new issue in science, because the godfathers of this endeavor, individuals such as Newton and Galileo, felt the basic creativity of the universe came from God, and that their job was to more fully apprehend the nature of his divine plan. In that sense, nature was like scripture, the word of God—only a more accurate rendering, not subject to the inevitable errors of human translation. God was, as Newton dubbed him, the "divine legislature" prescribing laws to the cosmos he had crafted. It was a mechanistic universe, and science's job was to understand the workings of the mechanism.

Indeed, French physicist Pierre-Simon Laplace was so emboldened with the discoveries of science in the eighteenth century that he claimed he could predict the future of the entire universe. All he needed, he explained, was to know the exact properties, positions, and forces acting on every atom in the universe *at any one moment*. Such knowledge would allow him, according to the laws of physics and chemistry, to predict the entire future—every last thing right down to what happens at the final moment of our universe. If this were a universe ruled entirely by eighteenth-century mechanistic laws of physics and chemistry, this speculation actually makes sense. But the hubris of his thought experiment was based upon a fundamental error—the idea that the universe functioned like a massive clock whose gears, once set in motion, had an inevitable outcome. We might call it a universe for type-A people—no novelty, no spontaneity.

Today, that particular conception of our universe is in tatters. This cosmos is much more complex, indeterminate, and creative than Laplace ever dreamed. No longer can we observe and report on nature's static truths; scientists today are digging much deeper, probing to discover not just how nature is but how it has become what it is. Such knowledge provides clues to the secrets of nature's past creativity, but also hints at how it might be transforming into something entirely unpredictable in the near and distant future.

"So what *is* the source of creativity in the universe?" I ask Bloom directly.

For once, he doesn't tell me a story, or set a long context with his reply. He answers simply with three of science's most important words: "We don't know."

CHAOS, COMPLEXITY, AND CREATIVITY

Leaving the Bloom brownstone well after midnight, my mind was buzzing with questions. Bloom does not have answers to the story of cosmic creativity; no one does. But his colorful love of nature's remarkable history had helped me appreciate just how much there was left to understand about Mother Nature's laboratory of novelty.

We had finished the evening by discussing how human consciousness developed in the evolutionary process. While there are many examples of the kind of creative leaps Bloom described, consciousness certainly qualifies as Exhibit A on the how-the-heck-did-evolution-come-up-with-*that*? list. You don't have to spend too much time at conferences where people are trying to explain human consciousness before you understand the appeal of the "God created it" point of view. But the larger point is that there are some real mysteries still to be solved in our understanding of evolution and many of them have a lot to do with this notion of creativity and novelty. Unfortunately, as Bloom mentioned, as soon as you start invoking the word "mystery," people tend to get nervous and think that the next thing you're going to do is to start invoking supernatural forces and ancient omnipotent deities to explain it.

One of the most popular new ways to think about the creation of novelty is to invoke the discoveries of the new complexity sciences. These new fields suggest that there are common principles, perhaps even universal laws, that apply across all kinds of "complex adaptive systems." A complex adaptive system can be anything from a collection of molecules to a software program to an economy to a human being. If we could understand these principles, so the story goes, we could perhaps explain a lot of the higher forms of order, complexity, and novelty that we see in the natural world. There is a relationship between novelty and complexity, these theorists point out. As complexity increases, new and higher forms of both novelty and order emerge that mark evolution's advance.

Perhaps the most important insight to come out of the complexity sciences is the idea that novelty is produced at the edge of chaos. That means that novelty generally *doesn't* appear in systems that have too much order and stability. It takes chaos, instability, and greater degrees of freedom to produce the conditions for higher forms of order and novelty to occur. A system generally has to be thrown into disequilibrium for new and higher levels of self-organization to emerge. We don't need to have degrees in physics to understand this subtle but profound truth. Here, again, is a principle that applies across many scientific disciplines and also to that most interesting of nature's species, human beings. Consider what it takes to get a human being to change his or her life, to really make entirely new breakthroughs, to achieve novelty, to actually evolve. Absent outside pressures, it is simply very rare for an adult human being to fundamentally change. People tend toward inertia and homeostasis, just like any complex adaptive system. But if you apply a little pressure from life—say, a health crisis, an economic crisis, or an emotional challenge—then the dynamics of the system change. Suddenly that human being is thrown into some subtle or gross form of disequilibrium. It could even be a positive challenge—a career promotion, or an unexpected windfall. All of these will disturb the human system to such a degree that new behaviors, new leaps of positive evolution, or higher forms of order may come into existence. Life may self-organize at a new level. So it is right at the edge of chaos that the system has its greatest potential for change, its ripest moment for evolution.

In the case of the human, we have to include the reality of conscious choice, which plays a role in the process, but the principle is the same. Of course, it's also possible to go too far. Too much chaos doesn't lead to higher order, just more chaos and disorder. It creates not breakthrough but breakdown. One can understand from this example why former *Wired* magazine editor Kevin Kelly writes, "The art of evolution is the art of managing dynamic complexity." Again, it's a general principle that applies to the evolution of any complex adaptive system, whether it's a corporation, a computer network, or a human being.

Much of the interest in these new sciences has been driven by our growing appreciation for just how extraordinarily complex the universe that we live in is. Not only has the macrocosm of cosmology and astrophysics proven to be vaster than we ever imagined, and the microcosm of particle physics and quantum mechanics more extensive and boundless than we ever thought possible, but the sheer unbounded complexity of the world as revealed by science has also grown along with the explosion of our knowledge. A thundercloud might not immediately seem like the most complex thing to a casual observer—until you actually try to model and predict its behavior on a computer. Then one begins to appreciate just how daunting it is to understand the behavior of complex systems. In fact, much of this new science has been driven by the information age. The growth of computing power has opened up new vistas for our understanding of complexities once far beyond our reach.

There is more to the nature of complexity than merely understanding the behavior of systems like thunderstorms or a flock of birds or the movements of the stock market. What about the evolution of new forms, new systems, new structures, new life, entirely new species? What about Bloom's "material miracles"? Might complexity science have something to say about that? In fact, there seems to be an interesting connection between evolution, complexity, and novelty, one that we are just beginning to understand. And it starts with a simple but explosive truth: *As biological evolution proceeds, complexity increases.* And another truth: *As cultural evolution proceeds, complexity increases.** Complexity, it would seem, is essential to understand. In fact, we could probably just simplify the two statements and say: *As evolution proceeds, complexity increases.* It's a fascinating piece of data, but why is it true?

To be fair, it should be acknowledged that this is a controversial statement in evolutionary theory, in part because no one understands exactly why it is true, or can even come up with an agreed-upon measurement to determine *whether* it is true. Yes, we can see that the atomic composition of iron, which was formed in the core of stars, is more complex than hydrogen and helium atoms, which are the original (post–Big Bang) building blocks of the elements. But developing a quantitative definition of complexity that applies to physical systems is quite tricky. The same goes for cultural evolution. It seems rather obvious that our globalizing culture today is far more complex than previous civilizations, but it's one thing to know this intuitively, and quite another to come up with a clear measurement. (Are we three times more complex than the

---

* It might seem that the evolution of political arrangements defies this trend. After all, we noted in the previous chapter that they have, in some respects, proceeded toward more unity, not more diversity. But on close examination we see that greater integration and increasing complexity can easily go hand in hand. Autonomous political entities may be fewer, as we noted, but their underlying subcomponents are immensely more complex.

ancient Greeks? Ten times? A million?) As Robert Wright notes, "precisely defining complexity or organization is such a notoriously frustrating task that many people give up and fall back on an intuitive definition, like Supreme Court Justice Potter Stewart's famous definition of pornography: 'I know it when I see it.'"

In his 2003 book *Biocosm*, James Gardner explored the idea that there may be an as-yet-undiscovered law of nature that can help explain the evolution of complex systems. Gardner writes:

> How did the beautifully intricate and interdependent ballet of molecules through which DNA is replicated and proteins are assembled first get composed? How did the universal coding scheme of DNA, common to virtually all living creatures, arise in the first place? How did Mother Nature first assemble the lavish palette of options from which the grim reaper of natural selection could prune away all but the choicest samples? ... Should we be searching for rules of self-organization that precede the operation of the great Darwinian principle of natural selection? Is this self-ordering process altogether different from the process of random mutation, which is the only source of novelty condoned by orthodox evolutionists?

An attorney by training, Gardner is a superb writer who brings a legal sense of clarity to his second profession of science writer and complexity theorist. He notes that this conundrum has led certain theorists, notably complexity scientist Stuart Kauffman, to speculate as to the existence of a possible fourth law of thermodynamics, one based upon a new understanding of self-organizing complex systems. Kauffman is one of the leaders of the field of complexity science. He has spent his career investigating nature's remarkable tendency to exhibit a kind of spontaneous, self-organizing order—"order for free" he calls it—which can produce higher forms of organization and novelty in all kinds of systems and potentially speed evolution's advance. Some complexity theorists, Kauffman included, suggest that Darwin's natural selection might ultimately prove to be only one source of novelty, and perhaps not even the most important creative process in nature. Kauffman calls it the "powerful idea that order in biology does not come from natural selection alone but from a poorly understood marriage of self-organization and [natural] selection."

In fact, in his recent book, *Reinventing the Sacred*, Kauffman argues for a new way of looking at the universe that appreciates the "ceaseless creativity" of the cosmos, a creativity that he suggests is partially beyond *any* natural law. Kauffman feels that the innate capacity of the universe to produce extraordinary new and novel forms of order with emergent, unpredictable properties is so profound that we must rethink the reductionist worldview of science. Not everything is reducible to physics and chemistry and the interactions of physical systems, *not even in principle*. "My claim is not simply that we lack the sufficient knowledge or wisdom to predict the future evolution of the biosphere, economy, or human culture," he writes. "It is that these things are *inherently* beyond prediction. Not even the most powerful computer imaginable can make a compact description in advance of . . . these processes." Kauffman's book is a passionate argument for the open-ended, exuberant creativity of life and of evolution, a creativity that we can embrace in our lives even if we do not fully understand it through our science.

Kaufman's bold declarations about the inherent unpredictability of the natural world are reflected in the work of another well-known complexity theorist, Stephen Wolfram. A brilliant scientist by any measure, he is an expert on computational systems and has done much of his work on what are called "cellular automata." These are algorithms that take very simple computer programs, sets of rules governing cellular grids, and then run them over and over again to see what kind of interesting behavior emerges. For example, imagine a cellular grid on a computer screen. Now imagine a few basic rules that determine whether any given cell on the grid should be black or white, based on the color of its neighbors at any given moment. Then run a program that repeats those same instructions over and over again. As the grid moves through various cycles in which the same rule is applied, it goes through all kinds of changes. And what Wolfram noticed was that with very simple rules, one can produce complexity of a stunning degree. Novel levels of order emerge in the grid, self-organizing structures that would seem to be completely unpredictable from the instruction set. His conclusion is straightforward but far-reaching—simple axioms can lead to extraordinary novelty and complexity. It is a conclusion that has led him to speculate that perhaps the extraordinary novelty, order, and diversity we see in the natural world is at root nothing more than a set of axioms, a foundational cosmic instruction set, let loose to create this magnificent universe. In fact, he has speculated that even the universe itself might be little more than a giant cellular automaton computer. Perhaps a simple code is at the heart of it all—perhaps everything we see in nature, from the most sublime and beautiful falling leaf in the Amazon to the winner of the 2020 NBA title can all be traced back to a basic, understandable set of principles. Wolfram calls this discovery "one of the more important single discoveries in the whole history of theoretical science."

It is fascinating to consider there might be still undiscovered universal laws of nature which are, as Gardner puts it, "locked into the very logic of the universe and that endow cosmic processes and their constituent subroutines with an inherent tendency to produce cascading phenomena of increasing complexity." Can Wolfram's automata really produce the kind of category-leaps of emergence— for example, the appearance of life—that we find unfolding across evolution's long history? Bloom, for one, describes Wolfram's work as providing important evidence for exactly the kind of open-ended, metastructural axioms that he suggests are being reflected in the creative activity of the cosmos.

Such speculations, at this point, are as much philosophy as science, but not too long ago such thoughts would likely have been dismissed as overly religious or metaphysical. Such is the compelling mystery of complexity. It has inspired a generation of scientists to look with a different kind of lens at the vast universe and offered tantalizing hints that there may be a lot more creativity built into the hidden algorithms of evolution than is yet dreamt of by our most able philosophies.

INTELLIGENT DESIGN AND NOVELTY

Blaise Pascal once noted that humans are stuck between two infinities: the infinitely small and infinitely large. "We wander in a vast medium," he wrote, "always uncertain and drifting, pushed by one wind and then another." But Pascal was missing a critical third piece of the puzzle that changes those observations. As Teilhard noted, we are living expressions of a third infinity—the infinitely complex. We seem to be in the midst of a vast process of increasing complexification, one that has also given birth to human beings, a species that just happens to have something called the human brain—what some have called the most complex structure in the known universe.

It is easy to underappreciate the depth of this mystery. This universe hasn't gone from dust to more complex dust. It's gone from dust to the music of Mozart, from nondescript hydrogen to the heroes of Homeric poetry. But for some of the more religiously inclined, the complexity of nature hasn't just been a cause for deeper reflection about the nature of evolution; it has been a cause to doubt it altogether.

Remember, we live in a universe in which one of the defining characteristics, we are told by physicists, is the second law of thermodynamics, which states that the entropy of an isolated system will tend to increase over time. Simply put, that means that all other things being equal, things will tend toward disorganization, disorder, and decay. In fact, I've been conducting a regular experiment in the second law of thermodynamics while writing this book. Leave a hot cup of coffee on the desk next to the computer and don't disturb it for half an hour, and I can say without a doubt that it does tend to get cooler and less tasty over time as the heat dissipates into the air. To my occasional dismay, I've yet to discover an exception to this rule. And in fact, I'm not the only one. It's pretty much an accepted absolute law of the universe in all scientific quarters. Perhaps astrophysicist Sir Arthur Eddington put it most succinctly when he wrote, "If your theory is found to be against the second law of thermodynamics I can give you no hope; there is nothing for it but to collapse in deepest humiliation." It should also be said that there are those who quibble with the basic presumption of the second law in relationship to this universe, the issue of whether or not the universe is, in fact, an isolated system (I guess the thinking is that there may be other universes or a metaverse that are connected with ours) but we'll leave that aside for now.

At first glance, the evolution of complexity may seem to contradict this basic principle of entropy. After all, our biosphere hasn't fallen into a more disorganized and decayed state over the last few billion years, and even the universe itself doesn't exactly seem hell bent on a sort of null point of randomness, equilibrium, and homogeneity. Quite the opposite. It doesn't seem to be running down; it is just as arguably ramping up! So does the evolution of complexity over the course of biological history (and perhaps even cosmological history) mean that there is something wrong with the second law of thermodynamics? Questions like this are part of what has inspired scientists like Kauffman to search for a "fourth law."

I find this question particularly interesting because of an e-mail I received recently from my teenage nephew, who attends a private Christian high school in Houston, Texas. His education has been exemplary—except when it comes to evolution. His e-mail contained an attached file called "Evolution: Not a Chance!" that had been given to him by one of his teachers. It used the argument that because the second law of thermodynamics tells us that systems run down and things fall apart and that this is not what evolutionary theory tells us, therefore something must be wrong with evolutionary theory.

This argument is dead on arrival. Biological evolution gets around the implications of entropy because the Earth is not an isolated system. We are riding the light, so to speak, using the sun's energy to power our way along. The sun is losing energy, but we are the beneficiaries. Like a rogue user of someone else's powerful Wi-Fi outlet, the earth naturally uses energy from the sun to power the entropy-busting evolution of life. Remember, I explained in chapter 4 that evolution happens on the borders, on the energy gradients, at those places where energy flows from one system to another. And it's at that great border between earth and space, where the energy from the sun floods our planet's surface, that evolution has happened, and where the most extraordinary novelty has come into being.

"You have to hand it to the creationists. They have *evolved*," jokes Eugenie Scott, executive director of the National Center for Science Education in Oakland, California, which monitors attacks on the teaching of evolution. Joking aside, it's true that some religious opponents of evolution are getting more sophisticated in their attempts to discredit the accepted scientific paradigm. And some have tried to use our emerging recognition of the extraordinary complexity and creativity in evolution as ammunition in this battle. In the 1996 book *Darwin's Black Box*, biologist Michael Behe argued that certain biochemical systems in the body are so intricate and so interdependent—that there are so many independent systems and parts that must work together in order for these systems to function at all—we should consider them to be "irreducibly complex." He suggests that such systems simply could not have evolved through the basic methods of natural selection. The absence of any one of the parts in these systems would have caused the whole system to cease functioning, the argument goes; therefore, in order to be adaptable, these systems would have had to evolve all at once in some great leap forward, as opposed to evolving more gradually in the stepwise accumulations that characterize Darwinian evolution. In his book *No Free Lunch: Why Specified Complexity Cannot Be Purchased without Intelligence*, William Dembski explains Behe's argument:

> A system performing a given basic function is *irreducibly complex* if it includes a set of well-matched, mutually interacting, nonarbitrarily individuated parts such that each part in the set is indispensable to maintaining the system's basic, and therefore original, function. The set of these indispensable parts is known as the *irreducible core* of the system.

Despite being embraced by many prominent politicians, Behe's and Dembski's ideas, usually called intelligent design in the media, have received little love from the scientific community and have even been heavily criticized by more forward-looking religious thinkers. The problem with Behe's and Dembski's argument is not merely that it often seems to be such a transparent attempt to make a place for an outside-the-process intelligent designer, something akin to the traditional Christian God. Rather, it is that they have such a limited view of what God's creativity might look like. I would suggest that the larger problem with intelligent design is not that it has questions about the capability of current evolutionary theory. That is a natural part of scientific inquiry and reasonable minds can render judgment on the veracity of those arguments. My problem is that many tend to follow such criticism by implicitly or explicitly appealing to the most uncreative picture of nature available, the "God just created it" picture of the universe.

I find it ironic that what many cite as wondrous evidence that evolution is remarkably creative, others use as a reason for retreat from and rejection of evolution altogether, and appeal to an omnipotent designer. Even if we are spiritually or religiously inclined, why would we want to fall back to a vision of God as omnipotent designer rather than a God whose work is only more fully revealed in the creative impulse of the evolutionary process? As theologian John Haught wrote in a response to Behe:

> What I object to is the narrowness of any theological approach that seeks to defend the idea of God, or to understand God's relationship to an evolving universe, and especially the evolution of life, by focusing exclusively on "design." It is not surprising that such an approach leads many of its proponents to reject evolutionary science or to edit it severely. Design, as Bergson pointed out long ago, is unrepresentative of what we now know about the strange story of life on this planet. And today it fails to advance dialogue between theology and biological science.
>
> Writing as a theologian, my point is that we should not abstract, and then isolate, the element of order from the often disturbing fact of novelty in actually living phenomena. Our understanding of God is considerably diminished by failing to reflect fully on the fact of novelty in nature. The concept of "design" is too stiff to accommodate either the complexity of nature or the depth of religious experience of God.

I would elaborate on Haught's excellent points with the observation that not only is our understanding of God diminished by failing to fully reflect on the fact of novelty, but our understanding of ourselves is diminished as well. As our picture of the universe continues to expand and we grow more and more cognizant of the creative power of nature, it is simultaneously as if the creative capacity we once reserved entirely for God has seemed to flow out of heaven and into earth. And as a product of nature's creation, we share in that bounty. As our picture of evolution grows more creative, so does our picture of ourselves. God's onetime omnipotence has become our own creative potential. And there is an interesting feedback mechanism at work here as well. As creativity more deeply informs our collective sense of self and our worldview, our capacity to consciously and creatively influence the evolutionary process itself grows. Intelligent life-forms become partners in and creative contributors to evolution's "design" rather than just the passive result of its mechanics. This may enhance the power and potential of the process itself in ways that we cannot even begin to foresee today. Some have suggested that this feedback mechanism may in fact be a critical part of the process. In an *EnlightenNext* interview, Jim Gardner compared cosmological evolution to the development of an embryo. He explained to me that when an embryo begins to develop, at some point along the way it needs critical feedback to complete the developmental process:

> When an embryo begins to develop, every step in that development is not specified in advance by the DNA sequence. What happens is that the embryonic development reaches stage one, and then the tissue complex—that is, the embryo—starts sending signals back into the DNA, which modulate further expressions of the gene into new tissue. So it's a feedback loop, and the informational complexity inheres in that feedback process, not simply in the nucleotide sequence. That's truly the extraordinary miracle of it. The process of embryogenesis is exquisitely programmed to actually take account of the state of its own ongoing development and to use the succeeding stages of development as a sort of augmentation to the basic instruction manual, which is the DNA contained in the genome.

Might humans, or intelligent life in whatever form, play that same role in the cosmological, universal evolutionary scheme of development? Might we in some way represent this feedback loop for the universe itself? Could our reflection on the evolutionary process itself be an essential element not only in fulfilling the next stage of our own development but in creating the next novel stage of cosmogenesis? Gardner's hypothesis is one of the most original—and compelling—evolutionary speculations that I have come across in some time.

Whatever we ultimately realize about the connection (if any) between the destiny of intelligent life and the destiny of the universe, there is at least one thing that I have become convinced of beyond a reasonable doubt: Creativity and novelty are not simply curious sidebars in the evolutionary script, beautiful by-products of a random cosmos or fortuitous flourishes of a designer God. They are written into the very cosmic narrative itself. So if we want to build a worldview informed by the evolutionary dynamics of the universe, creativity must not be peripheral to our efforts. As we will see in the chapters to come, creativity cannot really be delineated as one subject among many. Its footprints run across all of these pages, and its significance is ultimately not just a scientific matter but a spiritual one as well.

CHAPTER SEVEN

Transhumanism: An Exponential Runway

*The human species can, if it wishes, transcend itself—not just
sporadically, an individual here in one way, an individual there
in another way, but in its entirety, as humanity. We need a
name for this new belief. Perhaps transhumanism will serve:
man remaining man, but transcending himself, by realizing
new possibilities of and for his human nature.*

—*Julian Huxley, "Transhumanism," 1957*

Meat is messy." That was one of the first things I heard as I arrived at the 2009 Singularity Summit in New York City. The
speaker on the stage was Anna Salamon, a researcher at the Singularity Institute for Artificial Intelligence, and she was addressing a
crowd of eight hundred or so (mostly men) on the subject of "Shaping the Intelligence Explosion."

The conference is about the future, but the setting was decidedly historic. We were sitting in the beautiful auditorium at the 92nd
Street Y, a building that has seen, since its founding in 1874, many

great figures pass through its doors—cultural icons such as poets T. S. Eliot and Dylan Thomas, dancer Martha Graham, and cellist Yo-Yo Ma; social and economic leaders such as Gloria Steinem and Bill Gates; and political figures such as Mikhail Gorbachev, Jimmy Carter, and Kofi Annan. On this day, it was playing host to a new species of intellectual known as the transhumanists, who see in the explosion of information technology a new and hopeful future on the near horizon, a future so different from what has come before that it has been dubbed "post-human" or "transhuman."

The term "transhumanism" was coined in an essay of the same name by evolutionist Julian Huxley in 1957. Huxley, whose humanist convictions ran deep, called for a new exploration of human nature and its possibilities rooted in our understanding of evolution. This new evolutionary adventure, he suggested, might best be labeled "transhumanism." But even Huxley might have raised an eyebrow at the statement I heard as I walked into the auditorium that day. *Meat is messy.* Its meaning, however, is revealing. "Meat," for those not schooled in cyberpunk slang, means simply the biological body and all of its strange and bothersome whims and weaknesses. The body is meat—squishy, biological stuff. In contrast to digital worlds and virtual realities, the body is relatively stationary, hard to change, easy to damage. In a word: *messy.* Sure, it might seem a crude term to describe something so intimate as our living, breathing bodies, but for those who see the future in the bits and bytes of digital information, the body, as currently constructed, is part of evolution's past. The future is altogether different—free, open-ended, and unconstrained by the physical and temporal limitations of the flesh.

William Gibson, science-fiction writer and hero to nerds everywhere, popularized "meat" as a term for the biological body. In the 1980s, his book *Neuromancer* was an instant classic, presaging the themes of the movie *The Matrix* by well over a decade, and Gibson became a prophet to the first tech-savvy generation in history.

In the book, the main character, Case, steals from his employers and they respond by destroying his ability to enter virtual reality, prompting this passage describing his fate:

> They damaged his nervous system with a wartime Russian mycotoxin. . . . The damage was minute, subtle, and utterly effective.
>
> For Case, who'd lived for the bodiless exultation of cyberspace, it was the Fall. In the bars he'd frequented as a cowboy hotshot, the elite stance involved a certain relaxed contempt for the flesh. The body was meat. Case fell into the prison of his own flesh.

Notice the religious language. Gibson's use of "the Fall" and his hero's "contempt for the flesh" calls to mind religious condemnations of bodily desires. Of course, religion itself is far from the mind of most transhumanists, many of whom are materialists to the core, but there is a religious flavor to their conviction that the march of technology is telling us something critically important—not just about human culture but about life, the universe, and the evolutionary tendencies of both. You will never meet people more passionate about, and in some cases concerned with, the possibilities of the future.

Looking around at the people gathered for this annual event, which featured top researchers and theorists in fields such as artificial intelligence, nanotechnology, biotechnology, robotics, life extension, genetics, space travel, and computational theory, the mood was clearly one of excitement and anticipation. As the next speaker took the stage and began to explain the ways in which science is attempting to create a whole-brain emulation that will allow us (theoretically) to transfer our consciousness out of the physical body and into other substrates, I noticed that many of my fellow attendees were paying close attention—but not to the speaker. Their eyes were transfixed by the screens on their laptops and mobile phones. No doubt the subject matter preselects for technophiles, but I've never seen so many people listen to a speaker while engrossed in the digital devices in front of them. If the future of evolution is about human-machine integration, about "the marriage of the born and the made," as Kevin Kelly puts it, then this audience was definitely breaking new ground. As soon as any of the speakers said something even remotely striking, a small army of tweeting techies would post it, discuss it, correct it, contextualize it, and even fact-check it online for all to see. In fact, I confess that soon I was engrossed in my own small screen (well, not my own rather dated handset but my more tech-savvy colleague's), finding it almost easier to follow the content of the presentations by reading the Twitter feed from the audience than by trying to keep up with the sometimes convoluted PowerPoint slides on the stage.

Some of the presentations were fascinating, like one on computer-controlled vehicles. Others were incomprehensible—one on quantum computing comes to mind. A discussion of the future implications of artificial intelligence was among the more thought-provoking sessions, but there were many others that seemed bizarrely abstract and futuristic to the point of absurdity. For example, one presenter went to great lengths in exploring the philosophical implications of replicating brains and transferring them to different bodies. What would the new John Doe think of the old John Doe whose brain he had just replicated?

I had come to this gathering of the digerati to reflect upon the evolutionary implications of life in the midst of the "Technium," as Kelly calls the sum total of all of our culture and technology. Increasingly, thoughtful observers of the information-technology revolutions of the last decades are noticing an important truth—these technological innovations, with all the tremendous promise and peril they bring to our lives, do not represent an *aberration* from human cultural evolution but rather an *intensification* of the process.

And this process, they argue, has been well under way for millennia. Some even claim that the basic principles we find in the midst of today's rapid technological change are not novel at all, but have informed the processes of biological evolution and cosmic evolution since the beginning of time. Whatever the case, it has long been clear to me that an evolutionary worldview that can genuinely help us to make meaning in the twenty-first century must also help us to contextualize the information revolution in the arc of the larger evolutionary story. It must somehow connect seemingly disparate things—like chemistry and consciousness, Darwin and digital technology, qualia and quantum computing, cellular automata and subtle layers of spirit.

THE REVENGE OF THE NERDS, SQUARED

So what exactly is the singularity? Well, some have called it the only hope for humanity's future. Others are convinced it's the natural culmination of biological and cultural evolution. I like to think of it as the ultimate revenge of the nerds.

The term "singularity" means different things to different people, but in the broadest sense it means the union of humans and machines, the born and the made. But more specifically, it refers to a time in the near future when an important technological threshold will be crossed. The exact timing of this event varies depending on the theorist. Some think of it as the moment when computers' raw processing power will exceed the processing power of the human brain. Others use it to refer to the moment when artificial intelligence will recognizably transcend human intelligence, and various measures are suggested for judging that milestone. But however one defines the term, the idea is that at some point in the not-too-distant future, change will be happening so fast that a threshold will be reached beyond which it will become increasingly difficult to recognize human culture as clearly "human." We will be augmenting and altering our bodies and lives to such a degree that we will begin to lose the sense of clear continuity with who we have been. Our tools of prediction, these theorists tell us, cannot account for this level of cultural dislocation and unpredictability, as fundamental categories of human life are challenged and changed by the accelerating powers of our emerging technologies. It will be, in the words of author Vernor Vinge, "an exponential runway beyond any hope of control."

If there is a hint of dystopia in that statement, don't be surprised. That is the reason why so many sci-fi novels of the last decades, including Gibson's, present a rather ambivalent vision of the future. After all, transhumanism really does mean that we're going to be *transcending* the long-established categories that make us human. We're going to be messing with our genetic code, altering our minds and memories with tiny nanocomputers, radically upgrading our sensory apparatus, dramatically extending our life spans, and according to some, eventually transcending the biological body altogether, not to mention creating artificial life and intelligence that may surpass or even supersede our own.

But while it's understandable that there is more than a little angst about the outcome of such radical experimentation, the mood among those in the audience in New York reflected the overall mood of the transhumanism movement: a powerful faith in the future. It is a conviction in the redemptive potential of technology and an optimistic belief that the more we embrace the coming changes, the more power we will have to positively shape the inevitable ups and downs of this "exponential runway" we are about to taxi down. After all, they tell us, our evolution is all about technology. It always has been, since the first human fashioned the first Stone Age tool millions of years ago. And so the technological changes coming down the pike aren't an aberration from nature; they *are* nature!

They are evolution in action. After all, we are *meant* to transcend ourselves. And now the technological means to our own transcendence are at hand. All aboard; our post-human destiny awaits. And there is no point in trying to stop the train, the transhumanists shout from the conductor's seat, for it has already left the station.

It is not a mystery why transhumanists get accused of being naïve and even dangerous utopians, playing God without the necessary wisdom or knowledge. And this criticism is only fueled by statements like one from AI researcher Hugo de Garis, who once told an interviewer: "The prospect of building godlike creatures fills me with a sense of religious awe that goes to the very depth of my soul and motivates me powerfully to continue, despite the possible horrible negative consequences."

But whatever naïveté they embody, whatever wisdom they lack, whatever values they are missing, the transhumanists are also keepers of a truth that most of humanity is ignoring. These technologies are coming. Genetics, robotics, nanotechnology—"GNR," as that triumvirate is often called. Sooner or later, they are coming, and they are going to change everything. I think of the transhumanist movement as a kind of evolutionary wake-up call, an early-warning system for a sleeping culture. The marriage of the born and the made *is* in our evolutionary future. In fact, increasingly, it is in our present. So how do we make meaning in that world? How do we take responsibility for the consequences of that world? Religion, as we've known it, can't answer those questions. Neither can science. Philosophy is struggling with them. And New Age paeans to the wisdom of indigenous cultures are certainly not going to help. Only a new kind of worldview could possibly meet the spiritual, moral, and philosophical demands placed on us by a post-singularity world, whenever and wherever such a cultural moment might appear.

It is not entirely clear who first used the term "singularity." Some have traced it back to the mathematical genius Jon von Neumann, who is said to have used it in the 1950s in conversations about the future of technology. In our own time, Vernor Vinge was perhaps the first to publish the term in the context of technological change. In a 1993 essay entitled "The Coming Technological Singularity," he predicted that humanity was "on the edge of change comparable to the rise of human life on Earth. The precise cause of this change is the imminent creation by technology of entities with greater than human intelligence."

In the last two decades, the person most associated with the term "singularity" has been futurist and inventor Ray Kurzweil. In his 2005 book *The Singularity Is Near*, Kurzweil made a powerful case for the notion that technological change is advancing at such a rate that we will soon reach a point where it will transform "every institution and aspect of human life, from sexuality to spirituality." Under Kurzweil's influence, the singularity has become not only a more popular concept but a more flexible one as well. These days, if one is interested in the singularity, it tends to mean that one is interested in all of the many ways—genetics, nanotechnology, robotics, artificial intelligence—that human-created technology is promising to revolutionize what it means to be human. "The singularity," Kurzweil writes, "will represent the culmination of the merger of our biological thinking and existence with our technology, resulting in a world that is still human but that transcends our biological roots."

For Kurzweil and other transhumanists, there is nothing new about our brave march toward the transcendence of biology. In fact, one could argue that humans have been trying to transcend their biology from the moment they became aware that there was such a thing. Starting with the prehistoric clubs used to augment the strength of our blows, every tool humans have ever used could justifiably be placed into this category. And yet our powers are increasing at a frightening rate, producing a world that is no longer changing lifetime by lifetime or generation by generation but year by year and even day by day. This technological change gives us the visceral sense that time has a directional arrow, that we are headed somewhere, that the future will be demonstrably different from the past, that history is not merely moving the pieces around on an already existing chessboard but is creating new games with new rules on entirely new playing fields.

In the 2000 documentary *No Maps for These Territories*, William Gibson reflects on the realities of our fast-changing technological lives, noting, "The truth-is-stranger-than-fiction factor keeps getting jacked up on us on a fairly regular, maybe even exponential, basis. I think that's something peculiar to our time. I don't think our grandparents had to live with that."

I couldn't agree more with Gibson, whose insights into the effects of technological change on human psychology always ring with authenticity. He is an ambivalent but prescient forward observer of the transhumanist parade. But the truly interesting word in Gibson's quote is "exponential." It is in the meaning of this term that we find the connection between the techno dreams of the transhumanists and the emerging insights of an evolutionary worldview. And it was Kurzweil who made the critical link—with one insight that is shaking up futurists, challenging technologists, and changing the way we think about the path of human evolution.

WARNING: THE FUTURE MAY BE CLOSER THAN IT APPEARS

Legend has it that Albert Einstein was once asked what the greatest power in the universe was. His answer? *Compound interest*. Now, whether or not the great physicist ever actually said these words is a question we'll leave for the historians, but the power of the insight should not be doubted. I found myself contemplating this thought recently as I sat in the lobby of Kurzweil Technologies in Wellesley, Massachusetts, waiting to speak with the man who put the term "singularity" on the lips of the intelligentsia.

The power of compound interest is based on a simple formula—that the interest obtained is added to the original principal, thereby becoming part of the calculation for the next iteration of interest. The amount accumulated doesn't just change by regularly adding a fixed amount to the original principal. Rather, it builds on its own momentum, so to speak. That is why all retirement planners beg twentysomethings to save money while they are young, even if only a little, because of the incredible advantages gained by those additional years of accelerating returns. This principle is key to what Kurzweil likes to call exponential growth—it accelerates as it moves forward. And it keeps accelerating . . . and accelerating.

As a young inventor, Kurzweil came across this principle in the process of trying to create timelines for his projects. "When I finished many of my projects, three or four years after the original idea, invariably the world was a different place," he explained. "Most inventions fail not because people can't get them to work, but because their timing is off. And so I became an ardent student of technology trends, and this interest has taken on a life of its own."

When I say Kurzweil is an inventor, I'm not kidding. His office is covered with the many national and international awards his creations have garnered over the years. And there are pictures of him with artists (Stevie Wonder is prominent), politicians (one has Bill Clinton shaking his hand), and luminaries of all kinds thanking him for his work. All of his inventions have the same theme—they showcase the power of new kinds of technology, often for humanitarian causes. For example, he is the inventor of the first text-to-speech synthesizer device that allows the blind to read, making him something of a hero to visually impaired people around the globe. He is also the creator of the ubiquitous Kurzweil keyboard, which allows electronic synthesizers to create sound that is indistinguishable from a grand piano as well as other orchestral instruments.

Kurzweil Technologies is Kurzweil's mother ship, a technology incubator that takes bright new ideas all the way from the depths of his intuition through the stages of prototype and experimentation and then finally to market, at which point they are generally spun off into separate companies. These include FatKat, a hedge fund based on innovative pattern-recognition software, and Ray and Terry's Longevity Products, a company that produces life-extension supplements. His offices are also the home of his technology team, who keep close tabs on the important technology trends of the day.

"I have a team of people who gather data, measure different aspects of different technologies, and then develop mathematical models of them," Kurzweil explained. "The most significant trend that this investigation has uncovered is that the pace of change is itself accelerating."

He points out that it was only in recent centuries that people began to realize that technology was even changing, or that there was such a thing as technology at all. The industrial revolution changed our perception, and today people relate to change as a constant. But for Kurzweil, change is not a constant at all. It is growing exponentially. "According to my models," he explains, "we're roughly doubling what I call the 'paradigm-shift rate,' which is the rate of progress, every decade. So that means that the twentieth century wasn't a hundred years of change at today's rate of change, because we've been speeding up. It was actually twenty years of change at today's rate of change. Exponential change is quite explosive, so in the next century we'll make about twenty thousand years of change at today's rate of progress—about a thousand times greater than the twentieth century, and that century was no slouch for change."

Did you get that? Twenty thousand years of change in the twenty-first century? Talking to Ray Kurzweil is a little like entering a reality-distortion field in which the normal perceptions of evolutionary change suddenly speed up dramatically. Again, *exponential* change is the critical concept here. There is a world of difference between linear change and exponential change. And the human mind, to the extent that it considers change at all, is wired to think in linear terms. Futurists fall into the same trap. They project the future based on a reasonable extrapolation from the present of linear growth over time. Makes perfect sense—it's just wrong. At least, according to this particular pied piper of the singularity.

"Most technology forecasts and forecasters ignore altogether this historical exponential view of technological progress," Kurzweil writes in *The Singularity Is Near.* "Indeed, almost everyone I meet has a linear view of the future. That's why people tend to overestimate what can be achieved in the short term (because we tend to leave out necessary details) but underestimate what can be achieved in the long term (because exponential growth is ignored)."

Perhaps the best part of Kurzweil's presentation of the significance of exponential change is his demeanor. He is totally unfazed by it all. Kurzweil could tell you about the most wild, outrageous prediction for the future with almost zero emotion. His voice keeps dropping to lower registers toward the end of his sentences, keeping everything very straight and free of hype. The message conveyed is: "I'm looking at data; others are going with their gut." The contrast between the dramatic nature of the message and the purposefully understated style of the presentation creates a cognitive dissonance in my mind—like Ben Stein reading Allen Ginsberg's "Howl." Some might find it off-putting, but there is something endearing about it as well. And as we talked, I found Kurzweil to be warm, curious, and even quite funny at times—albeit in a quiet, introverted way.

It's worth noting that he may very well be the first person in history to fully appreciate the profound distinction between exponential and linear growth, a distinction that many feel applies well beyond the evolution of computing technology. In fact, Kurzweil has reams of data, chart after chart after chart, showing how the evolution of any information technology follows an exponential curve. We've all heard the famous example of "Moore's Law," in which Gordon Moore, the founder of Intel, noted that the number of transistors that can fit on a microchip doubles every eighteen months. It is something of an article of faith now, and whole design and production cycles in Silicon Valley are based on it. But according to Kurzweil, Moore's Law is really a reflection of a much larger principle. In fact, he points out, the microchip paradigm was already the fifth generation of information technology, and exponential growth runs through every generation since 1890.

"It's amazing how smooth these graphs are," Kurzweil says. "Look at wireless communications from Morse code a century ago to 4G networks today—smooth exponential growth for a century. It's surprising how predictable it is, when you consider that what we're measuring is the overall output of human innovation, creativity, and competition. You'd think it would be unpredictable, and indeed, specific projects are, but the overall result is not." Kurzweil points out that these historical trends of exponential growth continued, even given the drastic effects of the two world wars, the Great Depression, and many other disastrous cultural events. None of them significantly altered the curve of change. It reminded me of what Robert Wright noted about cultural evolution—unpredictable in the micro, but clear progressive trends in the macro.

To listen to Kurzweil is to take a journey into a world of technological accomplishments that seem like magic, existing in the far-off future, and then to compress them into the next few decades, if not years. Take the energy crisis. According to Kurzweil, the future has answers. Solar panels combined with nanotechnology should do the trick. "Solar energy production is doubling every two years," he tells me. "And it's only eight doublings away from meeting a hundred percent of the world's energy needs. And we have ten thousand times more sunlight than we need. We're awash in energy." Water shortage? The future has answers. "We're awash in water—most of it is just dirty or salinated," he says. "But we can convert it into a usable form with new technologies." Food? "We can create food that is safe and inexpensive with no ecological impact using hydroponically grown plants and in-vitro cloned meat without animals. Even PETA endorses that idea." Medicine? How about robotic red blood cells that are a thousand times better at holding oxygen than our own red blood cells? "You could sit at the bottom of your pool for four hours without taking a breath, or take an Olympic sprint for fifteen minutes without breathing." Or perhaps you'd prefer some robotic white blood cells, which are "dramatically more powerful than our ordinary white blood cells. They can download software from the Internet and destroy any kind of pathogen." The list goes on and on.

One of Kurzweil's common examples for illustrating exponential growth is the Human Genome Project, the international scientific effort to sequence the human genome. Started in 1990, the project, which was supposed to last fifteen years, was initially met with strong skepticism. Critics felt it was a ridiculous goal; that it would take generations, if not longer, to fully accomplish the ambitious task. Halfway through the project, doubts remained. Only one percent of the genome had been successfully accounted for. Surely the endeavor was doomed to failure. But in Kurzweil's exponential universe, things were actually right on track. As he explained to me, with exponential growth "you start out doubling tiny numbers, and by the time you get to one percent you are seven doublings away from a hundred. The project finished ahead of schedule."

I was taken with Kurzweil's optimism, impressed by his data, and moved by his dedication to making technology serve human advancement. But I was particularly interested in how he saw all of his work in the context of an evolutionary narrative. "My thesis *is* a theory of evolution," he told me, outlining a whole perspective on evolution that incorporates the idea of exponential change. For Kurzweil, exponential growth is not some temporary aberration from our evolutionary trajectory; it is practically the defining principle right from the start. One of the ways in which evolution has proceeded on Earth, he explained, is that it has tended to evolve whole new "technological platforms" on which evolution could occur. Theorists call this the evolution of evolution. "As an evolutionary process evolves a capability," he explains, "it then adopts that capability as part of its methods for evolution. So the next stage goes more quickly. And the fruits of the next stage grow exponentially." Exhibit A for this principle is DNA. "Evolving that took a billion years, but then biological evolution adopted it, and has used it ever since. The next stage, the Cambrian explosion, went a hundred times faster." Kurzweil points out that people tend to think of the Cambrian period, when all the body plans of the animals evolved, as a very special period of creativity. But from his perspective, it isn't special. It was just the natural result of a new capability or "technology" becoming available to the process. Eventually, through a series of such exponentially faster stages, the process produced a species that could create technologies itself, which was a further exponential leap. "So human technology and cultural evolution is a continuation of the process that created the technology-creating species in the first place," Kurzweil concludes.

Given Kurzweil's background, it is hardly surprising to hear him describe the universe in such technological terms. However, it was fascinating to me, in talking to Kurzweil, that he also had a conviction in the deeply *spiritual* nature of the process. "In my mind," he told me, "evolution is a spiritual process." Of course, he came to that conclusion through a very logical deduction:

What happens during evolution? Entities get more complicated. They become more knowledgeable and more creative, more capable of higher levels of emotions, like love. What do we mean by the word God? God is an ideal meaning infinite levels of all of these qualities. All-knowing. Infinitely beautiful. Infinitely loving. And we notice that through evolution, entities move towards infinite levels, never really achieving them, staying finite

but exploding exponentially to become more and more knowing, more and more creative, more and more beautiful, more and
more loving, and so on—moving exponentially towards this
ideal of God but never really achieving it.

THE DESTINY OF THE EARTH AND THE COSMOS

Ray Kurzweil is certainly not the only thinker to point out that the
rapid spread of information technology is not an aberration from
evolution's trajectory but an integral part of it. Indeed, some say
that we are in the midst of the greatest leap forward in the proliferation of information since Gutenberg started his printing press in the
mid-fifteenth century, or even since the earliest writings in Egypt
and Sumer and the original cuneiform tablets. It does seem almost a
given that the emergence of our new digital landscape constitutes a
remarkable moment in cultural development. Kevin Kelly recently
compared the significance of our age, and the birth of the Internet
to that of the historical inflection point 2,500 years ago known as
the Axial Age, when four of the world's major religions and several
other influential philosophical systems were all born in the space of
a century.

"There is only one time in the history of each planet when its
inhabitants first wire up its innumerable parts to make one large Machine," Kelly writes, with the breadth of vision common to thinkers with transhumanist persuasions. "Later that Machine may run
faster, but there is only one time when it is born. You and I are alive
at this moment. We should marvel, but people alive at such times
usually don't."

While Kelly and Kurzweil have been critical of each other's
work and their timelines of the future vary, they do share the con-
tagious optimism of fellow techno-evolutionaries. And maybe, like
them, we should marvel. But perhaps we should also feel a hint of

caution. After all, there are, to use William Gibson's words, "no maps for these territories." There is no user's manual for our newly awake "Machine" as it grows in knowledge and power.

Indeed, what exactly is this "machine" we are creating? What is its ontological status? Is it conscious? Is it alive? Is it a new form of intelligence? This birth is an event that has truly launched a thousand philosophical questions, not to mention spiritual and existential quandaries—as we try to come to grips with the fast-unfolding implications of our own creation. Where exactly is this digital daemon headed in the years to come?

In the first part of the twentieth century, Teilhard de Chardin had his own prescient thoughts about the evolution of our collective intelligence. More than half a century before the formation of the Internet, he wrote, "No one can deny that a network . . . of economic and psychic affiliations is being woven at ever-increasing speed which envelops and constantly penetrates more deeply within each of us." Gibson added his visionary voice to the mix in 1984 when he wrote, again in his classic novel *Neuromancer*, of a future digital matrix that was a "consensual hallucination" with "rich fields of data" where one could observe "bright lattices of logic unfolding across that colorless void."

Today, it is in the transhumanists' milieu that we find some of the most interesting, provocative, and generally outrageous predictions of what might be coming our way in the years and centuries ahead. Indeed, there have been plenty of people to notice that our global communications networks, taken together, seem to be constituting a collection of connections that has eerie similarities to the structures that make up a human brain. So, is the Internet the equivalent of a global mind? And even more relevant, if it were becoming conscious, would we even know it? The speed at which this new global mind has been born understandably emboldens the prophets.

But even if we stop well short of declaring the Internet to be a global brain, something of evolutionary significance is happening in the marriage of technology, biology, and matter. When I reflect for a moment upon the power of my latest smart phone, it is as if the matter in the phone itself has come alive with intelligence and power. Is there an analogy to life? Consider the difference between the matter in a rock and the matter in a living organism. The matter in living organisms has achieved a sort of freedom, autonomy, mobility, and intelligence that nonliving matter never approaches. So, does the difference between a smart iPhone and a dumb rock amount to the same sort of evolutionary leap? For some transhumanists, this is the essence of the evolutionary process: waking up matter by infusing it with intelligence and information.

As usual, Kurzweil has a thoughtful and radical take on the idea. "In my mind," he told me, "we will ultimately saturate the whole of the matter and energy in our area of the universe with our intelligence, and then it will spread out at the fastest speed with which information can flow to the whole universe. Eventually the whole universe will essentially wake up. Ultimately all of what I call the dumb matter and energy in the universe will be transformed into sublimely intelligent matter and energy. That is the ultimate destiny of the universe."

Such statements may seem grandiose and speculative, but remember that Kurzweil has based much of his career on the well-informed extrapolation of current trends into the future, and he isn't willing to abandon ship just because the time frames are getting cosmic and the conclusions unconventional. Indeed, he has been one of the very few, with the possible exception of a rare physicist or spiritual visionary, to point out a fact about the future evolution of the universe that is both completely sensible and completely remarkable as well. He points out that most speculations about the future of the universe don't take into account the evolution of intelligent life. The majority of theorists completely ignore its influence. But that makes no sense. If you just consider our own evolution over the last ten thousand years, not to mention the evolution of whatever other forms of intelligent life may be out there, and then extrapolate that process forward several billion years, it would seem reasonable to suppose that we (meaning whatever form of intelligent life we have evolved into) may have advanced to such a degree in our intelligence, sophistication, and technology that we have a say in the cosmic destiny of the universe. It's quite a notion to wrap one's mind around, but as Kurzweil points out, in his typically matter-of-fact style, it is simply a reasonable conclusion to draw from the data:

> The implications of the Law of Accelerating Returns is that intelligence on Earth and in our solar system will vastly expand over time.
>
> The same can be said across the galaxy and throughout the universe. . . . So will the universe end in a big crunch, or in an infinite expansion of dead stars, or in some other manner? In my view, the primary issue is not the mass of the universe, or the possible existence of antigravity, or of Einstein's so-called cosmological constant. Rather, the fate of the universe is a decision yet to be made, one which we will intelligently consider when the time is right.

If we see intelligence as being merely the furthest result yet of a long, linear evolutionary process that will simply continue in a future trajectory, unperturbed by the consequences of its creations, then perhaps there is no reason to consider the role of intelligence in shaping the destiny of the Earth or the solar system, much less forces at a cosmic level. However, if intelligence is an emergent property in an exponential curve, another game-changer in evolution's long history of game-changing creations, then trusting the blind forces of physics to determine the future of the universe may be no more reliable than trusting that purely "natural" forces will determine the future shape of an alpine glacier. As we are learning, the power of intelligence, for better or worse, plays a powerful role in the latter, and believe it or not, may someday play a role in the former. The only difference is scale.

AN INFORMATION ONTOLOGY

In his 2009 book *The Nature of Technology: What It Is and How It Evolves*, economist W. Brian Arthur explores the parallels between biology and technology and comes to the conclusion that they are becoming more and more synonymous over time. He points out that as our technology gets more and more sophisticated, it is beginning to look less mechanistic and more organic, taking on many of the functions and properties that we associate with living organisms. And he also points out that the more sophisticated our understanding of biology is, the more we grasp the extraordinarily subtle and complex mechanisms that make up the processes of our biological lives, the more we appreciate the essentially mechanistic nature of all of the interacting parts and processes. He writes:

> Conceptually, at least, biology is becoming technology. And physically, technology is becoming biology. The two are starting to close in on each other, and indeed as we move deeper into genomics and nanotechnology, more than this, they are starting to intermingle.

Now, Arthur is both an economist and a complexity theorist, and so the idea that biology is ultimately just technology, that life is essentially a function of very complex material processes producing higher forms of order, is one he is naturally comfortable with. But whatever we ultimately conclude about the nature of life itself, the idea that biology and technology are growing closer and closer has some very important implications. How are they connected? Here we come across a conviction that is very important to understand when it comes to transhumanism: *Everything is information.*

In order to appreciate the insights of transhumanism, one has to understand just how critical a role information plays in the transhumanist understanding of reality. Spiritually inclined individuals may dismiss these techno-futurists as being materialists, but I've come to believe that this is an inaccurate characterization. They are neither materialists nor spiritualists; they are informationalists. They have an information ontology, we might say. Down at the foundations of reality, where some see spirit and some see matter, they see information. It's the building block of their worldview. As Kurzweil explained at one point during our conversation, "Living creatures are information. Biology is an information process." In Kurzweil's universe, what is ultimately evolving is not life or matter or beings or even consciousness but the complexity and sophistication of information.

Physicists have long tended to think of the universe in terms of the most complex machines of their day. Remember when the universe was like a clock? It makes sense that scientists today would think that it's a giant information processor. Our emerging understanding of the role of information in all kinds of physical, biological, and evolutionary processes represents a significant move out of the billiard-ball universe of science's past, where everything could be reduced to tiny particles that collect together and crash into one another. And because information as a concept is much more closely related to the idea of intelligence, when we talk about the evolution of information processing it naturally places a higher premium on the role of intelligence in the evolutionary process. Was human-level intelligence an inevitable, or at least likely, outcome of the evolutionary process? In an information-laden universe, the answer would surely have to be yes.

An information ontology gets even more rich when we consider the work of philosopher David Chalmers, founder of the aforemen-

tioned Tucson consciousness conference, a maverick but highly respected academic who has introduced a theory of consciousness that
places information in a central role and suggests that both matter
and subjective experience are "double aspects" of information. He
suggests a "conception of the world on which information is truly
fundamental, and on which it has two basic aspects, corresponding
to the physical and the phenomenal features of the world." In other
words, Chalmers suggests that both consciousness (phenomenal)
and matter (physical) are, in some sense, the result of a world built
out of information. It is no accident that when I queried him, Kurzweil cited Chalmers as his favorite philosopher.

My point is not to endorse Chalmers's view, or to say that it
represents the exact way in which Kurzweil and other transhumanists see the evolutionary universe. Rather, it is to show that there
are powerful ideas at the leading edge of science and philosophy
that take us beyond the easy dualisms of the past and confound the
polarized categorizations of science, technology, and spirituality.
They encourage scientists to not dismiss consciousness as the concern of feeble-minded romantics. And they encourage the spiritually inclined to resist their Luddite urges and the all-too-common
association of the mechanisms of technology with a cold, indifferent
material universe. An information-rich view of evolution need not
be reductive or spiritless. Even theologians have noted this truth,
pointing out that if God or Spirit works in mysterious ways, one
of those mysteries might have to do with that hard-to-categorize
power of information in the evolutionary process. "The quiet, unobtrusive way in which information insinuates itself into the chemistry of life," writes evolutionary theologian John Haught, "serves to
demonstrate that there can be a kind of influence operative in nature
that is not reducible to sheer material force."

LIVE LONG ENOUGH TO LIVE FOREVER

There are some rather interesting implications that flow from a worldview based on information. The first, and perhaps most significant, is captured in the title of one of Kurzweil's books, *Live Long Enough to Live Forever*. Consider this: If the essence of what makes a human being is neither the physical body or brain, nor some immaterial soul, but *information*, then it would theoretically be possible to remove a human being from the physical substrate—remove the human software, in other words, the patterns of information that make up the self, from the physical hardware. Indeed, perhaps it would be possible to move a human being from body to body without doing fundamental damage to that person—damage such as, well, death.

"Ultimately, we can and will transcend our biological limitations," says Kurzweil in his usual deadpan voice. And it must be said that he has put his money and body where his mouth is. He takes a couple of hundred supplements every day, has essentially cured himself of diabetes, and has embarked on an extraordinary health regimen, which he has chronicled in several of his books on the subject. The hope—and it is a hope he shares with many transhumanists, including English scientist Aubrey de Grey—is that we can discover ways to radically slow down the aging process in order to "cure aging," as de Grey explains it. The timeline they are looking at is years and decades, not centuries. It won't be long at all before we have people living to 120, maybe 150. And given exponential growth, who knows what the term "senior citizen" will mean once the singularity kicks in. "The first thousand-year-old is probably less than twenty years younger than the first one-hundred-and-fifty-year-old," de Grey says.

Kurzweil believes that by the 2030s and 2040s we will already be well on our way to transcending the many limitations of our biological bodies. "There is not a single organ of the body that is not being augmented already. It's just at a fairly early stage. People say,

'Oh, I like my biological body,' but we're not going to be limited to one biological body; we'll have multiple biological bodies." This prediction depends on more than exponential growth; it depends on a metaphysical leap of faith. Yes, if the essence of Carter Phipps or the essence of Ray Kurzweil is information and the kind of information that can be captured by human technology, then it could actually make sense, as outrageous as it sounds. If our identity can be stored in bits and bytes of highly organized data, then surely it can be moved around into different biological substrates. But let's not mince words—that is a giant, massive *if*. It's the kind of sliver of possibility, however, that has already inspired a great deal of enthusiasm. Indeed, this scenario has been referred to as "the rapture of the nerds." And so whether or not we biologically "cure aging" in the long term is a moot point to singularity enthusiasts. What matters is that we survive long enough for technology to allow us to upload ourselves into immortality, that we live long enough, as they say, to live forever in the "bodiless exultation of cyberspace." After all, *meat is messy*.

My colleagues and I have often remarked over the years that one of the by-products of adopting an evolutionary worldview is the temptation to embrace the idea of immortality. We have particularly noticed this tendency among spiritual Evolutionaries, and have speculated that there is something about the power of a deep evolutionary optimism, a genuine sense of almost incredible possibility and potential at the level of consciousness, that can lead to the misguided conclusion that evolution is destined (in the near future) to give consciousness power over matter itself. We see this in many of the esoteric Western spiritual traditions, where the "light body" is often bandied about as a sort of ultimate spiritual goal—the idea that the highest levels of evolution involve a transfiguration of the flesh. Similar ideas can be found at the esoteric edges of Christianity and the yogic tradition of Hinduism. Immortality also makes an appearance in the evolutionary philosophy of Sri Aurobindo, whose work was in turn influential in Michael Murphy's focus on new physical capabilities appearing in the human species, a subject he carefully researched and investigated in his impressive book, *The Future of the Body* (though Murphy makes no claims to immortality). Clearly, there are less credible and more credible expressions of this basic impulse.

So it is perhaps not so unexpected to see the subject of immortality turn up on the technological side of the evolution revolution as well. Does that mean we should discount all such speculations as baseless fantasy? Admittedly, prolonging our biological life indefinitely still seems far-fetched to me, and dramatically transforming our bodies with our minds even more so. But in a world where life spans are ever-increasing and the science is advancing day to day, far be it from me to establish any definitive limit on what's possible. We shouldn't forget that we have added thirty years to our life expectancy in the course of the last century. But whatever we ultimately conclude about such notions in science or spirit, the most striking thing is to note the common source—evolution. To embrace the power and potential of an evolutionary worldview, and to embrace the faith in the future that it represents—whether that faith is associated with technology or consciousness—is to embrace a future with radically expanded limits.

Uploading our consciousness to computers is certainly a wild science-fiction scenario to speculate about over a good Bordeaux (after all, components of wine have been shown to have potent anti-aging properties). But we can appreciate the evolutionary challenges and potentials of these new technologies and their ambiguous effects on our humanity without assuming such radical scenarios. I often find it humorous how much intense concern there is in the singularity community over the eventual impact of completely unproven, nonexistent technologies. Indeed, transhumanists are particularly good at engaging in hypothetical moral dilemmas that may be getting a bit ahead of the technology. Hugo de Garis is a poster child of this tendency, as he has stirred up considerable consternation that we are heading toward an inevitable "artilect war" before the end of the twenty-first century between those who embrace artificial intelligence (cosmists) and those who reject it (terrans). In fact, he dedicated a whole book, *The Artilect War*, to the subject. It is a war that will kill billions, he insists. I appreciate the insistence that we face up to the coming technological changes with some measure of honest concern, foresight, and deliberation, but there is also a limit to how much emotional energy we should invest in hypothetical moral dilemmas that depend entirely on unknown technologies that themselves depend entirely on metaphysical leaps of faith! In other words, don't lose sleep over the coming artilect war . . . for now.

Nevertheless, the transhumanists have several important pieces of this evolutionary worldview we are forming. They have embraced the material, or informational, side of evolution's advance with a vengeance. They are making an important case that human-created technology, with all its wondrous promise and dangerous potential, is not an aberration from nature but an essential expression of nature's handiwork. Yes, they may take that point to its logical extremes, but if we are to form a worldview that does not retreat from the future, which can face the reality of life on the edge of some sort of singularity head-on, whenever it may come and whatever it might look like, then we cannot hide from the future and the consequences of our technological revolutions. To again paraphrase Stewart Brand, humanity is already playing God, and we had better get good at it.

Becoming good at playing God, I suspect, will mean a much deeper and more profound understanding of the evolution of human culture, human values, and ultimately human consciousness itself. Indeed, if there is an Achilles heel to the transhumanist movement, I would say it is their tendency to oversimplify the nature of mind and consciousness, and to over-conflate consciousness with informational complexity. It is an ontological sleight of hand that, once employed, allows a whole host of imagined outcomes that might otherwise be difficult or even off-limits. But I will also say that in my conversations with Kurzweil, he expressed an authentic, open-minded interest in the subject of consciousness. Of course, consciousness is a difficult subject for any theorist, and Evolutionaries are no exception. And so, having explored the outer limits of material and technological evolution, it is to this inner dimension of life that we now turn.

PART III
RECONTEXTUALIZING CULTURE

CHAPTER EIGHT

The Internal Universe

O, what a world of unseen visions and heard silences, this insubstantial country of the mind! . . . An introcosm that is more myself than anything I can find in a mirror. This consciousness that is myself of selves, that is everything, and yet is nothing at all—what is it? And where did it come from? And why?

—*Julian Jaynes,* The Origin of Consciousness
in the Breakdown of the Bicameral Mind

Gong . . . Gong . . . Gong. The church bells rang out through the crisp morning air on the top of the sacred mountain. The sun had just risen over the distinctive jagged cliffs of Montserrat, Spain, and this venerable alarm clock heralded the arrival of the new day, calling the pilgrims to work and worship. But I needed no rousing. The bells seemed to ring inside my body, reverberating back and forth within my consciousness. I was sitting upright, perfectly still, meditating quietly as the first rays of sun fell on the ancient streets. I was in the middle of a ten-day retreat being held next to the cathedral at this pilgrimage site.

There is nothing like doing nothing to make one appreciate the mystery of consciousness. A soundless thought appeared in my mind like a water bubble rising in the vast ocean. Slowly, silently, it drifted away, faded into the distance, and the inner ocean was again calm. And for a moment, "ocean" was the right word, as my interior world seemed vast, spacious, all-encompassing. In fact, it didn't seem to be inside me at all; I seemed to be inside *it*.

I wasn't alone in my quiet reverie. All around the large hall were men and women of various ages and backgrounds, sitting in quiet attention, their focus turned inward, their bodies still. Meditating each day with two hundred others, one discovers a unique kind of camaraderie and intimacy, one that has nothing to do with the diversion of idle talk, the clash and confluence of personalities, or the intensity of shared emotions. In those hours of stillness, boundaries and barriers begin to fade, and it can feel temporarily as if two hundred people are sharing one meditative field, a singular interior consciousness. In the deepest meditations of that retreat, I can remember thoughts arising, and for a moment, having no idea if a particular bubble in this vast internal sea was actually my thought or someone else's.

Later each day, as the sun would make its journey down the backside of this "serrated" mountain, I would hike upward, following the steep paths to the high walkways that ringed the upper peaks. Here and there, small temples and shrines, one-time shelters for reclusive monks, dotted the stark but beautiful landscape. Religious pilgrims had been contemplating God on these hillsides since the twelfth century. On those rugged slopes, overlooking the Catalan countryside, I would think about consciousness.

Questions as to the origins and meaning of consciousness seem to be increasingly common today. It is as if more and more people on both sides of the science and spirit fence are trying to upgrade our understanding of what this inner dimension is all about. Scientists seem to have noticed that there is nothing in the existing account of our origin story that easily explains the mysteries that lie in the mind of the human. As Robert Wright has pointed out, "While I think natural selection provides a satisfactory account for [evolution], I do think there is still one massive mystery, and that is why consciousness, or sentience, exists at all, why there is subjective experience. And I don't think many evolutionary biologists appreciate the depth of that mystery." It is that remarkable mystery that has inspired a new generation of theorists and researchers to rev up their fMRI machines and CT scans at labs around the country, hoping that better scans of the material brain will give us hints as to the workings of the mind. But even as science begins to ask questions once considered immaterial, both literally and figuratively, something else is happening as well. Many spiritually inclined theorists are questioning previous assumptions and looking past the magical, mythical, and mystical shrouds that so long have obscured this fundamental subject.

Sages and mystics for centuries have claimed that the inner dimension of consciousness is, in fact, more real than even the most tangible physical objects of the external world, and past philosophical movements, going all the way back to the early Greeks, have certainly entertained all kinds of notions as to what might be the significance of human self-reflective awareness. But more often than not, theologians have reflected on the divine origins of consciousness, scientists on the material origins of the brain, and philosophers have sort of bounced uncomfortably between them.

As I walked along the narrow mountain pathways, a strong wind whipped across my face, carrying with it a fast-moving fog that little by little encircled the peaks and raced down the hillsides. I thought about Henri Bergson's statement that consciousness is the "motive principle of evolution." Just as I could see the fog racing down the face of the mountain, but could not see the true source of its mobility, the wind that carried it forward, so Bergson thought that life and matter are, like the fog, borne up by the current of consciousness itself, caught in the great evolutionary blast of this vital but invisible creative force. To conclude that matter is the sole evolutionary engine, he suggested, is to make the mistake of assuming that clouds somehow are moving by themselves, when in fact, clouds, air currents, and condensation are all integral to one system, some of it seen, some of it unseen.

In the silence of retreat and meditation, it is this unseen dimension that moves to the forefront of awareness. One becomes aware of just how profound it is that human beings possess this interior space, this inner world. Consciousness becomes less a taken-for-granted backdrop to the ongoing drama of life and more a living presence, rich with an expanded sense of depth, meaning, and significance. We live in a culture focused largely on external image and surface concerns, and for that moment at least, as I walked the rocky pathways where Benedictine monks had once wandered in search of spiritual solitude, it was a relief to have the freedom to give the interior dimensions the attention they well deserve.

"The space you discover in meditation is not just a quiet place inside your own head," remarked Andrew Cohen, the spiritual teacher who was leading this retreat, the evening before. "It's a dimension of reality itself—a dimension of the cosmos. It's the *interior* of the cosmos. The interior of the cosmos is not the inside of that mountain—that's still the exterior dimension. The interior of the cosmos is *your experience of consciousness*. The exterior of the cosmos is matter. It's what we see all around us, and what we can see if we look through a telescope—apparently we can see back to the earliest beginnings of our material universe. But the interior dimension is consciousness. The cosmos is not just 'out there'; it's 'in here.'"

What Cohen was sharing with us was a relatively new way of looking at the experience of consciousness, one that cuts through some of the philosophical complexity surrounding the issue and connects it to the reality of an evolving world. Whereas ancient mystic seers may have meditated on the inner depths of formless, timeless consciousness-without-an-object and declared that "only *that* is real" and contemporary scientists may prefer to contemplate the more prosaic pathways of matter and declare that only *that* is real, I prefer to take a "both/and" approach to this ancient conundrum. Yes, the interior, subjective world is quite real. Consciousness exists as a living truth of our own inner experience, but it is not some alternate substance operating completely independent of the physical universe, as Descartes once thought. Rather, it is the *interior* dimension of the cosmos, a nonphysical side of the physical coin, so to speak. The interior and the exterior, consciousness and matter, cannot be separated, this perspective suggests. This internal cosmos, this world of awareness, subjectivity, perceptions, ideas, emotions, and so on, is not an unimportant sideshow in the cosmic drama, a simple shadow effect of the neurons in our brain, but a legitimate part of reality, knit into the ontological fabric of a universe (or multiverse?) that may yet prove more remarkable than even the most wide-eyed theorists over in the physics department have yet surmised.

In truth, "consciousness" is probably a clumsy word for this internal universe, a blanket term that tends to paint with one brushstroke a canvas that has much more subtlety to it. For example, there is what has been called "pure consciousness" or the experience of awareness itself, the consciousness of mystical meditation, also referred to as "consciousness without an object," "primordial consciousness," as Buddhists have called it, or the "ground of being," a term popularized by theologian Paul Tillich.* And then there are internal thoughts, psychological structures, emotions, values, convictions, etc., that we also associate with this internal world. And that's just the beginning. If the Eskimos, as legend tells, have twenty words for "snow," perhaps one day, in a more enlightened age, we'll have twenty names that will more accurately distinguish the different dimensions of the internal universe. But for now, one will have to suffice.

---

* This term is a translation from the German *Wesengrund*, and was used by both Husserl and Heidegger prior to Tillich. Tillich took the term from Heidegger's *Sein und Zeit*.

An evolutionary worldview offers at least two critical insights when it comes to consciousness. First, *consciousness evolves*.* The internal universe, like almost all other aspects of reality when seen from an evolutionary perspective, is not static. It is not fixed. It is not set in stone, either by an Abrahamic God or a genetic code. Evolution is happening in this dimension as well. The self develops and evolves. Human consciousness develops and evolves. Evolution, in this way of looking at things, cannot be reduced to the physical world, to the changing synapses deep in the structures of the brain. It is also occurring deep in the interior dimensions of our subjective lives.

The second insight is perhaps even more important. It's not just our personal subjective universe that evolves. It's also our *shared* internal lives, our collective interior. As I experienced on that retreat, consciousness is not merely a private affair, a personal, sealed inner container of subjectivity. That is a false way to think about this interior dimension. To some extent, we share the internal universe with others. When I am meditating on a beautiful mountaintop in Spain with two hundred others, we are not just experiencing two hundred separate inner worlds. We are participating in a collective field of consciousness—not a subjective but an *inter*-subjective experience. And this intersubjective space, which can become more heightened in a setting like that retreat, is in fact a dimension that exists all the time, independent of any particular experience.

Put most simply, this intersubjective dimension is *culture*. Often when we use the term "culture," we think of its many outward *expressions*—music, art, fashion, or social and political institutions. But another way to think about culture is to see it as a dimension that exists *inside* of *us*—the interior of the collective. Author Jean Houston describes it as the "living tissue of shared experience." It is where

---

* Technically, there may be certain types of consciousness that don't evolve. Most theorists suggest that pure primordial consciousness or the "ground of being" doesn't itself evolve. But every other object or structure found in the internal cosmos would be subject to the evolutionary process.

meaning, values, and agreements live—a real, internal world that is, I will propose, part of the evolutionary dynamics of the universe. We have spent a fair amount of time in this book discussing the topic of worldviews—their reality, their influence, and their importance for understanding culture. Recognizing the existence of this intersubjective dimension allows us to take that understanding deeper—to see the actual "place" in which worldviews form and develop. After all, a worldview is a collection of shared values, beliefs, and agreements, and where do these cultural constellations live if not in the inner space between us?

In our materialist society, where many people have a hard time even acknowledging the legitimacy of subjective consciousness, much less the reality of this relatively new concept called *intersubjective* consciousness, such assertions require us to step out of our usual patterns of thinking. They ask us to embrace the possibility that there may be more going on beneath the surface of society, in the subterranean corridors of our collective consciousness, than we previously realized. As we shall see, there are few distinctions in this book that are more important or more useful when it comes to appreciating the new perspective of an evolutionary worldview.

In this book, up till now, we've been primarily exploring the more conventional manifestations of the evolutionary process—the development of technology, social complexity, cooperative arrangements, novelty, and so on. In the chapters that follow, I'll be traveling off the beaten path and examining what evolution looks like when we turn our attention to the interior domains of consciousness and culture. Currently, this is more the domain of philosophy than science, but as we will see, the evolutionary principles that are being revealed in this inner dimension are remarkably congruent with those being mapped by scientists in the exterior world. As guides in this intangible territory, I will be drawing on the work of a number of individuals who have been central to ongoing efforts to re-contextualize both consciousness and culture in an evolutionary light.

PHILOSOPHY AND THE NOOSPHERE

When I returned that year from my Spanish mountaintop sojourn, I was refreshed and inspired. Without a doubt, my spiritual immersion in the interior world of consciousness had deepened considerably during my retreat, but little did I know that my own philosophical understanding of the terrain was about to take a leap forward as well. The catalyst for that leap was the manuscript of a book I had recently received—*Integral Consciousness and the Future of Evolution* by Steve McIntosh, a man who would prove to be an astute guide to the dynamics of what he called "the internal universe."

McIntosh is the founder of Now & Zen, a small business in Boulder, Colorado, that has been a modest success selling highly crafted "Zen Alarm Clocks" for the last couple of decades. Originally educated as an attorney, he attended the highly prestigious University of Virginia, where he trained his mind in the ways of the law and contemplated the works of the great founder of the University, Thomas Jefferson. There, along the beautiful walkways that Jefferson himself had helped design, McIntosh considered the great achievements of our founding fathers. He gained a deeper appreciation for how they had taken the new worldview of modernism that was just beginning to become established in eighteenth-century culture and applied it to governance, helping to create the first nation built from the ground up on Enlightenment philosophy. Decades later, it was his continuing interest in politics that led to our initial conversations—I was researching an article about his efforts to generate interest in a new kind of global governance. As soon as we got on the phone together, the effect was like a spark plug firing. There was an instant connection that would turn brief phone calls into multihour-long philosophical explorations. And it was in those conversations that I began to understand more directly the relationship between ephemeral things like consciousness and practical things like politics.

"We might say that every problem in the world today is first and foremost a problem of consciousness," McIntosh explained to me in our first call, "and every solution involves the raising of consciousness." Now, don't get the wrong idea. By "the raising of consciousness," McIntosh didn't mean sitting around campfires in encounter groups exploring our feelings. No, he was talking about the actual evolution of the "internal universe," as he called it, the development of our shared values and worldviews. After all, so many of the problems of our world are, as we have discussed, problems of worldviews—of the clashes between them and the limitations imposed by them. These powerful systems of culture inevitably condition, for good or ill, our outlook on life. If there were a way to more accurately understand how and why certain worldviews form, the relationships between them, and the dynamics of their internal structure, that knowledge would be invaluable in positively influencing cultural evolution around the globe.

Like many of his generation, McIntosh had firsthand experience of cultural change. He had grown up in Los Angeles in the wake of the 1960s, when what is often referred to as the postmodern worldview was just flowering in the culture. He hung out on Venice Beach and had "countercultural loyalties," but he was more interested in turning on and tuning in than in dropping out. Whip smart and motivated, he was interested in how the tremendous excitement and inspiration generated by the countercultural movements of the time would translate into real social progress. As the revolutionary energy of the '60s faded into the individualism of the '70s and the Aquarian Age became the New Age, McIntosh pursued the spiritual side of the movement with great vigor, becoming interested in the emerging intersection between science and spirit—"which has been a foundation for my intellectual interests ever since," he says.

But over the next decades, McIntosh began to see that amid all of the excitement of the countercultural movement, problems were becoming obvious as well. The original promise of a broad-minded, inclusive, cross-cultural, intellectually rich, science-friendly spiritual tradition had been buried under a mountain of esoteric ideas, fly-by-night creeds, and have-it-your-own-way philosophies, leaving many dissatisfied and turning elsewhere for spiritual sustenance. And moreover, the "anything goes" mentality of the movement was making it increasingly fragmented and ineffective as a force for social progress. All of which made McIntosh think more deeply about how culture actually evolves. What is authentic cultural evolution? What makes it sustainable and lasting? What makes it spiritually rich— good, beautiful, and true? What was the difference, for example, between the incredible philosophical flights of classical Greece, which were ultimately unsustainable by the culture of the time, and the inspired rationality of the European Enlightenment, which was able to take lasting root, developing into a rich new worldview that transformed every institution of society, changing human culture permanently, and eventually allowing a southern aristocrat named Thomas Jefferson to rewrite the rules of politics and jump-start a new kind of nation?

There was another influence in McIntosh's life as well, one that, as we have seen, shows up in just about every Evolutionary's life at some point—Teilhard de Chardin. Here was a thinker who embraced science with a passion but was rooted in the spiritual and religious dimension of life. And McIntosh was particularly struck by Teilhard's use of an unusual word—*noosphere*.

"Noosphere" is a critical term when it comes to understanding the relationship between consciousness and evolution. In fact, "noosphere" and "intersubjectivity," as we shall see, are quite related. They both point to that underappreciated, and in some cases unrecognized, dimension of evolution—the interior of the collective. "This idea allowed us to see cultural evolution from a spiritual perspective, and also from a scientifically informed perspective," recalled McIntosh.

"Noosphere" was Teilhard's word, though he wasn't the first

to use it in publication. That distinction goes to Russian scientist Vladimir Vernadsky, who appropriated the term from the Jesuit. It is an evocative idea and has inspired many cultural leaders today, from Mario Cuomo to Al Gore to Marshall McLuhan. The word is a composite of the "sphere" part of atmosphere, lithosphere, stratosphere, biosphere, and so on, and the term "noetic," which means the realm of knowledge. Teilhard used the term "noosphere" to describe what he referred to as the "thinking layer" that surrounds the planet like a thin, invisible envelope of collective consciousness, representing the sum total of humanity's interior life. Just as evolution first formed the biosphere, a planet-enfolding sheath of living organisms, the noosphere, according to Teilhard, represents the next staging ground for evolution's advance. And as cultural evolution progresses, the noosphere itself grows more dense and complex, more rich and intense, more focused and multilayered, with all the interior qualities—good and bad—of an evolving society. Cultural evolution happens right here, in the collective interior life of humanity. "We have to realize that evolution has leapt beyond the biological context," McIntosh explains.

For Teilhard, there was always a relationship between the evolution of interior consciousness and the evolution of exterior complexity. "The Physical and the Psychic, the Without and the Within, Matter and Consciousness, are all . . . functionally linked in one tangible process," he wrote. For example, the human brain, looked at from the outside, is one of the most complex things we know of in the entire universe. Looked at from the inside, it also houses the most advanced form of consciousness we know of in the entire universe. Such observations give weight to the idea that there is an important relationship between the evolution of physical complexity on the outside and the evolution of consciousness within. Teilhard called it the law of complexity and consciousness. And for him, it applied not just to physical organisms but to large collectives as well. According to this law, as the noosphere progresses, we should see a correlation between the complexity of our techno-socio-economic systems and the sophistication of our collective consciousness. And as we have discussed, many point to the evolution of communications and information technology as demonstrating this very trend and marvel at Teilhard's prescience.

The birth of the noosphere, according to Teilhard, coincided with the birth of *Homo sapiens sapiens*, the birth of self-reflective thought in higher primates, the birth of the animal who knows that he knows and has an interior life unlike any the world had ever seen. Science today cannot say exactly when this threshold of self-consciousness was crossed, or even if threshold is the right word. Was it a sudden and solitary moment, a dramatic and immediate breakthrough? Or a gradual process? All we know for sure is that 40,000 to 60,000 years ago, something quite dramatic happened in the evolution of the hominid mind. As psychologist Robert Godwin puts it, "existence mysteriously becomes experience." During that phase of history, early humans passed an evolutionary tipping point. Art, creativity, language, self-identity, and new technologies all burst onto the scene, a creative cultural explosion that was both unprecedented and unexpected.

"Why, all of a sudden, do humans all over the globe begin expressing an urge to create, to bring into being beautiful artifacts that serve no utilitarian purpose?" asks Godwin. For Teilhard, at least, the answer had to do with a very specific kind of breakthrough that was achieved during this time period. As our brains became larger and larger, they retained the capacity for increasing consciousness, interiority, subjectivity, and psychic awareness. And slowly (though rather quickly in evolutionary time), as the first higher primates made their journey out of Africa into uncharted lands and the new mental heights of Homo erectus and Australopithecus and Neanderthal man, consciousness was intensifying. The complexity of interior lives was increasing. Finally, in one particular primate, consciousness reached a culmination. The rising temperature reached the boiling point. We don't know the exact conditions, or reasons. But we do know that something unprecedented happened. Teilhard evokes this moment in *The Human Phenomenon*:

> [E]verywhere the active phyletic lines grow warm with consciousness towards the summit. But in one well-marked region at the heart of the mammals, where the most powerful brains ever made by nature are to be found, they become red hot. And right at the heart of that glow burns a point of incandescence.
>
> We must not lose sight of that line crimsoned by the dawn. After thousands of years rising below the horizon, a flame bursts forth at a strictly localised point.
>
> Thought is born.

Brian Swimme, in *The Hidden Heart of the Cosmos*, tries to imagine the moment when conscious self-awareness first appeared in what he calls the Earth Community, to picture "a pioneering hominid at the dawn of human history" awakening to the awesome fact of awareness. "Some animal entered into the experience without understanding what was taking place," he speculates, "for no other animal had ever been in that mode of existence before." Godwin envisions "a luminous fissure . . . the dawning of an internal horizon in a universe now divided against itself, the unimaginable opening of a window on the world."

Whether or not these characterizations are correct in their representation of the events of 40,000 years ago is secondary to the fact that we have yet to fully appreciate the implications of the miraculous evolutionary leap that took place in that time period. Teilhard describes it as the birth of another world. "Abstraction, logic, reasoned choice and inventions, mathematics, art, calculations of space and time, anxieties and dreams of love—all these activities of inner life are nothing else than the effervescence of the newly formed center as it explodes on to itself," he writes. And he goes on

to make the critical distinction that human self-reflective consciousness, when compared to the animal consciousness that preceded it,
"is not merely a change of degree, but a change of nature, resulting
from a change of state."

Teilhard was not the only early-twentieth-century thinker who
began to suspect that there might be a realm within the collective
mind of humanity that mysteriously connected one person's inner
psychic space to another's. In fact, the great Swiss psychologist Carl
Jung coined the term "collective unconscious" to explain a similar
idea. Jung began to recognize that his patients were coming across
images, symbols, and psychological forces in their own inner lives
that he suspected were connected to a deeper layer of the human experience, archetypal currents that transcended the personal psyche
of any one individual and his or her history.

This realization prompted Jung to suggest that beyond the subjective psyche, there is a collective dimension of the interior life that
we are all connected to, and he termed that interior realm the "collective unconscious." He wrote that there exists a "psychic system
of a collective, universal, and impersonal nature which is identical
in all individuals." Now, at first glance Jung's collective unconscious and Teilhard's noosphere may seem like different terms. After all, they were meant to describe different types of phenomena.
One is explicitly evolutionary; one is not. One has psychological
overtones; one has scientific overtones. But once you get past the
superficial differences, it's hard to avoid the conclusion that these
two great theorists were, each in his own way and according to
his own language, shouting from the mountaintop about the exact same breakthrough—that there is a *collective interior* dimension
to consciousness. Whether we call it the noosphere, the collective
unconscious, or intersubjectivity, it is real. It is an integral part of
the evolutionary dynamics of nature, and I suspect it will help revolutionize our understanding of cultural evolution in the coming
century.

NOT I, NOT IT, BUT WE

While the term "intersubjective" may be unfamiliar, what it points to is not at all so. In fact, we experience the intersubjective dimension so often that we take it for granted. But think about it for a minute. If consciousness is entirely subjective—contained in the individual brain—then how could we relate to one another so intimately, so profoundly, so deeply? As Ken Wilber puts it, "How on earth do you get in my mind, and I get in your mind, enough that we are in each other to the point that we both agree we can see what the other sees? However this happens, it is a miracle."

McIntosh, whose writings on the subject have been greatly inspired by Wilber's philosophy, writes that "Relationships exist in the internal space 'in between us,' not wholly in our minds and not wholly in the minds of those with whom we are related, but mutually inside both of our minds and often simultaneously. These relationship structures are partially independent from our individual subjective consciousness, but at the same time internal and invisible."

When two people share a political conviction, when two lovers share an intimate moment, when two colleagues build a business together, they are each, in some small way, creating, participating in, and sharing a real, intersubjective world space. We could say that those relationships have an independent reality in the internal universe of the noosphere. Spend time with any long-married couple and you can really appreciate this. Whatever the personalities of the individuals in the relationship, the relationship has its own life, its own structure, its own complexities. It is almost an entity unto itself.

"Relationships have an ontological reality that is not just in my head," McIntosh explained to me in one of our conversations. "If I have a relationship with a six-foot rabbit named Harvey, that's a completely subjective reality within my brain. But if you and I are becoming friends, there's an independent reality of our relationship, which takes form and has a systemic existence. And even though the relationship exists in the overlap of our mutual experience, we can also begin to recognize that this relationship itself has a structure. It's what I would characterize as a system of culture."

"People think that because they can't see it, culture doesn't exist," says Grant McCracken, an anthropologist who has made his name convincing corporate leaders to take notice of this dimension. Culture, he tells us, is "the meanings and rules with which we understand and act in the world. This makes culture sound amorphous and absurdly abstract, I know. But let's put this another way. Culture is the very knowledge and scripts we will someday build into robots to make them socially sentient creatures. At the moment, we're still teaching them to climb stairs." McCracken's work points to a truth connected to this dimension that corporate leaders struggle with every day. Corporations, like relationships, have their own subculture, an independent collection of values and attitudes that define a company and influence it greatly. Change a few people, even a few hundred, and the culture doesn't necessarily change at all. Changing corporate culture is hard, any consultant will tell you, because you're not just dealing with individuals but the collective momentum of this larger cultural structure. Yes, it was certainly created by the intent of many autonomous individuals but once established, it can seem to act as an organism unto itself—an independent, intersubjective system—one that can exert influence and resist change.

"Of all the dimensions that are easy to miss, intersubjectivity is at or near the top," writes Wilber in *Integral Spirituality*. Our modern scientific efforts have for the most part heavily focused on the external, physical, objective universe. Science has struggled to explain, through the limited lens of biology, how these tool-making bipeds have been able to develop such a remarkably sophisticated culture in so little time. Indeed, looking for an explanation for cultural evolution by appealing solely to biology reminds me of that joke about a guy searching for his keys under the streetlamp. "Where did you lose your keys?" a helpful stranger asks. "Over by my car," the man replies. "Then why are you searching under the streetlamp?" asks the confused stranger. "Because this is where the light is," the man answers. Similarly, science has tended to stay away from consciousness because of the murky and confusing nature of the subject, preferring the more well-lit domains of biology and physics. But if we want to fully understand the nature of evolution, and especially cultural evolution, consciousness is where the keys are.

Remember, however, that the kind of consciousness we are talking about here transcends the lone individual's subjective interior. The war between science and spirit is often framed as a war between subjectivity and objectivity, between the private, personal world of the individual and the external world of nature, between the "I" and the "it." But there is another realm, which is distinct from both of those. And most people do indeed miss it altogether. But they miss it in much the same way people "missed" gravity before Newton. It's hiding in plain sight.

To help contextualize this new terrain of inquiry, McIntosh likes to use an example from history: the way we understood the human body before and after the Renaissance. He points out that the interior of the body used to be off limits to exploration; it was considered to be almost mystical and there was a great deal of superstition and uncertainty associated with what lies underneath our biological skin. Remember, Michelangelo had to break all kinds of Church rules and cultural taboos in order to perform the dissections necessary to understand the muscular structures that allowed him to so marvelously depict the human form. And through his and others' efforts, we now know that there is nothing particularly mystical going on inside our bodies at all, just a set of extraordinarily complex systems and structures that we are still exploring and defining today.

The same, arguably, is true of this intersubjective dimension. It may seem almost mystical today, but I suspect that further exploration will reveal an interior universe that is complex and subtle yet readily yields itself to deeper understanding. In fact, it is already beginning to do so. And just as we can look at the external universe and see all of Darwin's "endless forms most beautiful" that have evolved in our biosphere, the internal universe—the noosphere, the intersubjective dimension—has its own complex systems, structures, and forms of consciousness that are evolving. Just as we have come to appreciate that the physical world is composed of a truly startling degree of complexity, that even the seemingly simplest structure is composed of complex configurations of smaller and smaller networks of particles and energy systems, so too we would do well not to underestimate the complexity of the internal universe. Every thought, every feeling, every reactionary emotion and each complex vision, careful calculation, or intuitive perception is built upon vast networks of interlocking and interdependent thoughts, implicit and explicit conclusions, values, perceptions, agreements, paradoxical perspectives, complicated decision-making processes, amalgams of images and ideas, psychological complexes, archetypal patterns, and multiple layers of awareness. It is truly a vast internal universe, and the recognition of this helps explain why the physical correlates of this dimension—human brains—are the most complex entities, as far as we know, in the entire world.

Can an exploration of this internal universe really be compared to the scientific exploration of the physical body? For some, talk of inner space inevitably borders on the spiritual. And to some extent I understand that. It certainly blurs the lines between science and spirit. There are a couple of reasons, however, why I am always hesitant to automatically associate this interior dimension with the spiritual. First, there are all kinds of thoughts and values that exist in the noosphere that aren't particularly spiritual at all. But perhaps the other reason is that people tend to associate the word "spiritual" with things that are inherently beyond rationality, beyond clear language, things we can only speak about in stories and metaphors. How many extremely bright, sensible people have I seen resort to the most strange, esoteric, nonsensical language when they begin to speak about the so-called spiritual dimension of life? They throw out their common sense altogether and surrender their levelheadedness—as if spirituality inevitably implied vagueness, silence, or irrationality.

Of course, there is nothing wrong with poetry, myth, and metaphor when it comes to this interior world. That's a time-honored way of representing it. And mystical insight does certainly involve transrational, nonconceptual states of being. But there is a huge difference between genuinely transrational spiritual states, intuitions, and experiences, and areas of knowledge that may be subtle, complex, and relatively unexplored yet hardly unnatural or unknowable. We should never blindly conflate nonphysical with supernatural. I suspect that over time we will discover that the science of this internal universe, while subtle in terms of our present conceptions, is well within the framework of a sensible, comprehensible universe. And just as the emerging understanding of the body in the Renaissance radically changed our capacity to enhance physical health, so too will a new understanding of the structures and systems that constitute the interior of human culture enhance our capacity to affect the health of our global society.

In fact, it would be wrong to give the impression that this internal universe is entirely unexplored. We investigate aspects of this interior world all the time, in sociology, anthropology, hermeneutics, and so on. Psychologists, of course, have been exploring this area with great intensity over the past century and depth psychologists even more so—Jung being just one example. Mystics and visionaries have pursued the spiritual depths of the interior universe for millennia. More recently, certain analytic philosophers with a focus on linguistics have directly explored the nature of symbolic communication between people as constituting essential elements of our social fabric. And that's just a small list among the many people who are exploring—directly or indirectly, explicitly or implicitly—this interior space. But most have done so without the kind of philosophical framework I am pointing to here. Like pioneers in a new land, they are each following particular trails and uncovering different characteristics of the terrain, yet none has a map of the whole, a sense of the enormity of the territory and the integrated pattern of its many dimensions. They also tend to be captured by the spell of solidity, lacking a rich sense of how our shared collective interiors are moving and changing, and playing a significant role in the dynamics of cultural evolution.

One dimension of this discussion that does feature prominently in public discourse is the idea of worldviews. But while our public intellectuals use the term in discussions of politics and globalization, they do so primarily from a surface viewpoint—noticing the patterns of external symptoms but lacking a sense of the underlying systemic causes that an in-depth perspective on consciousness and culture helps to provide. Worldviews may be more than just a label we apply to the common characteristics, values, and beliefs in any given group of people that our social scientists have noticed in their surveys. That objective data speaks to deeper intersubjective truths. Perhaps these worldviews have a kind of systemic, independent existence that must be taken into account if we are to understand how culture evolves, and how culture influences individual and collective perspectives on the world. In other words, perhaps they are quite real—not material, but real nonetheless.

So if this interior cultural dimension is real, then what does it mean for it to evolve? Indeed, *what* is actually evolving when it comes to human culture? McIntosh's answer is that it is "the quality and quantity of connections between people, taking the form of shared meanings, experiences, and agreements."

"Agreement" is an important word in understanding this new terrain. Returning to a biological metaphor, agreements are, as McIntosh puts it, "the cells of the noosphere." In this regard, he draws on the work of Jürgen Habermas, the German philosopher who made linguistic agreements so central to his work. So what is an agreement? For example, agreements as to the meaning of certain sounds and symbols, which we call language, are some of the most foundational to human civilization. Without these agreements, we are unable to communicate, much less form a cohesive culture. More complex agreements would be ideas of what's right and what's wrong, for instance, or religious beliefs like the notion that a transcendent God is keeping track of each individual's actions in some heavenly log book and will weigh up their fates accordingly on Judgment Day. It is only the tacit agreement of individuals that gives such a notion its cultural power. Those outside the Judeo-Christian tradition do not lose sleep over their record in the Book of Life.

We can think of these agreements or "cultural cells" as the building blocks of worldviews. It is the accumulation of agreements into larger and larger constellations that eventually results in massive and complex internal structures or worldviews, composed ultimately of thousands and thousands of tiny agreements, just as an organism is composed of thousands of tiny cells. Some of these agreements will be foundational to the worldview and some more cosmetic; some will have to do with the deep, underlying logic of that worldview, and some the surface behavioral manifestations. But perhaps the hardest part to grasp is that these internal structures, once formed, have a reality that exists independently of any particular individual, a momentum of their own. And it is the recognition of this reality that allows us to treat these complex worldviews as legitimate internal structures of evolutionary unfolding that abide by clear principles and natural laws.

A GAME-CHANGING PERSPECTIVE

The evolution of worldviews is something that's always much easier to see in retrospect. For example, it's easier to see, with a bird's-eye view of history, how the evolution of the Axial Age religions, beginning in the fifth century BC, was likely a natural evolutionary response to the brutality of the Iron Age culture of Persia and the Middle East. The clarity of hindsight reveals how the modern worldview of the Western Enlightenment was a natural evolutionary response to the excesses, corruption, and failings of the traditional Christian worldview of the Middle Ages. And from the distance of decades it's not difficult to appreciate how the countercultural movements of the nineteenth and twentieth centuries, culminating in the seismic shifts of the 1960s, were a reaction to the many problems created by the excesses of modernity and the industrial revolution. But Evolutionaries are not just interested in the past; they are interested in the future. And charting the *next* cultural emergence takes much more than a rich sense of history. It takes a deeper knowledge of how evolution operates in the internal universe of values, agreements, and worldviews . . . and a little talent for informed speculation. "If you'll pardon the comparison, you and I are like beatniks sitting in a café in North Beach in 1952 trying to imagine what the sixties are going to be like," McIntosh said to me over dinner one evening at a conference in California. "We have an inkling of it, but we don't have the cultural structures to point to yet."

The first time McIntosh began to get an inkling of what a new worldview might look like, he told me, was in 1999, just as winter was beginning to descend on the slopes of the Colorado Rocky Mountains. Ken Wilber, who had also made his home on the edge of the Rockies and who had done so much to bring highly sophisticated evolutionary thinking to the spiritual side of the countercultural movement, had just formed Integral Institute, and was hosting gatherings of sympathetic thinkers, authors, scholars, and spiritual leaders in Boulder. Soon McIntosh's name was on the invitation list, and he entered into a new sphere of relationships informed by new values and agreements. He began to recognize that this nexus of interacting ideas, people, and philosophies represented not just a blip on the cultural radar but a genuinely new movement, supported by an original philosophical framework that, in the words of Wilber, transcended and included many of the problems of the countercultural movement.

Like the coming of the dawn in a darkened landscape, McIntosh's own vague intuitions of the problems with the countercultural movement were illuminated. He was able to see its great strengths—a respect for spirituality, a tolerance of other cultures and faiths, a capacity to appreciate the perspectives of others, an intense concern for the environment, a newfound respect for indigenous cultures, a deep compassion for the plight of the victims of society, and a passion for minority rights. At the same time, he could see now, more clearly than ever, the many weaknesses of this cultural movement—its "anything goes" individualism, its tendency toward narcissism, its pathological resistance to all hierarchies, its social idealism combined with political impotence, and its dangerous propensity to romanticize premodern cultures, peoples, and faiths. He could see all of these things from the inside out. After all, the counterculture was his culture. He had shared its agreements and lived its values—explored its limits, indulged its opportunities, suffered its disappointments and embodied its dreams. But now he was changing and beginning to appreciate a powerful new truth—that the postmodern, countercultural worldview was just that, a worldview. It wasn't the be-all and end-all of cultural evolution. "I was just filled with this spirit of excitement," he explained to me. "I was beginning to see with more clarity than ever that worldviews, these structures of culture, had an evolutionary reality. They had an existence that was independent of any particular person's writing or thinking."

As the conversations between McIntosh and me deepened, I felt a similar spirit of excitement. As I began to contemplate the evolutionary reality of these structures in the internal universe, I knew this was one of the most potentially game-changing facets of an evolutionary worldview when it comes to facilitating authentic social and cultural transformation.

Consider the great philosophers of the Enlightenment in the eighteenth century. They did more than simply change the institutions of society; they helped reinvent the underlying agreements and values that society held to be important. Eventually, those new agreements called for new rules, new laws, new rulers, and entirely new institutions. And today, we are still unraveling the contours of the modern worldview they helped bring to life. They created, we might say, new intersubjective agreements, and eventually an entirely new worldview, a new way of making meaning, and in so doing, changed culture in a deep and lasting way. The problems of their time were not just institutional or material or economic or political; they were also problems of consciousness, and so they sought to raise the consciousness of European culture. They built new institutions that reflected this change in values.

In the same way, if we are to both evolve our own culture and understand the evolution of cultures in general, we will need to more deeply understand how agreements form worldviews. This book is, from a certain point of view, an attempt to shed light on the kinds of agreements that are informing, and creating, an evolutionary worldview. Furthermore, if we are to help other cultures in our globalizing world to move forward, to transcend the entrenched patterns that create political and social dysfunction, we will need to be able to see through the surface institutions and conflicts to the deeper structures that lie beneath, and learn how to enable them to mature. There is cultural leverage in this perspective, a way that we can effect change "from the inside out," so to speak, which means evolving the underlying agreements, values, and worldviews that influence both our personal consciousness and the collective institutions of our society. "Understanding the anatomy of an agreement," McIntosh predicts, "will be one of the new sciences of the noosphere."

While I don't expect the new sciences of the noosphere to be making their debut in mainstream university curricula anytime soon, and consciousness is still a term that understandably evokes more associations with personal meditation and private contemplation than intersubjective inquiry and shared cultural evolution, there are significant changes in the intellectual climate today that are opening up the interior world to objective analysis as never before. Once esoteric dimensions are being made transparent to our questing eyes, revealing new truths and shedding new light on existing ones. Biological development, we are learning, is just one of the tricks up evolution's sleeve. Once upon a time, evolution took a step, a leap, and catapulted itself into a whole new category. Evolution itself evolved. It created a new space in which to operate. We might say that human culture, in all its beauty and complexity, is the result. Biology and culture, genes and memes, can neither be perfectly separated nor grossly conflated. They are two distinct and influential parts of evolution's layered emergence. Consciousness, it turns out, is not a mere sideshow in the evolution revolution. It is an integral part of a comprehensive understanding of Darwin's favorite idea.

Chapter Nine

Evolving Consciousness: The Inside Story

If we have learned anything from the critique of reason that was initiated by Kant and completed by Hegel, it is that the very categories of our understanding are historically conditioned. We have come to recognize that there are mental structures that determine how we understand the world and ourselves and that these structures evolve in history.

—*David Owen*, Between Reason and History

A couple of years ago, while attending the "Towards a Science of Consciousness" conference in Tucson, Arizona, I happened to wander into a panel session that included Dutch theorist Jan Sleutels. The subject was the "evolution of consciousness" and his particular presentation was titled "Recent Changes in the Structure of Consciousness." Sleutels started out his talk by announcing that he was going to expose one of the central fallacies in our understanding of consciousness—the assumption that it hasn't changed in the course of human history. He called this assumption "the Flintstones fallacy"—the erroneous idea that human consciousness has essentially been the same for thousands of years, even as technology has developed significantly. We tend to imagine that humans in earlier ages were like the characters on the cartoon show *The Flintstones*—more or less like us, only dressed in more primitive fashions and carrying clubs instead of cell phones. The outer world may be evolving, so this fallacy goes, but the inner remains essentially the same. "Consciousness is generally seen as an endogenous asset of the mind/brain that is responsive to pressures on an evolutionary time scale, but that is largely unaffected by cultural history," Sleutels declared. "Substantial changes in recent history are ruled out a priori."

Sleutels then suggested what he called a more baffling possibility: that as recently as hundreds of years ago we might find "minds substantially different from ours, yet belonging to human beings *very* much like us. So much like ourselves, in fact, that we find it almost impossible to believe that our ordinary mentalistic vocabulary of beliefs and desires should *not* apply to them as literally as it applies to us."

The Flintstones fallacy is, I suspect, one of the great unexamined assumptions of our time. And as Sleutels suggested on that hot Arizona day, it is a fallacy that underpins a great many core assumptions we make about human history. It's not just the Flintstones—movies, historical novels, and even scholarly studies regularly portray our ancestors as having essentially the same emotional and psychological repertoire that we have today. How many of us have deeply considered the possibility that human consciousness may have changed significantly in the last five thousand years? Five hundred years? Two hundred years? Again, we come back to that overarching insight with which I began this book—the touchstone proposition of an evolutionary worldview. *We are moving.* That which we thought was fixed, static, and relatively unchanging is revealed to be fluid, moving, changing, and malleable. But it's easier to break the spell of solidity when we are looking at an external artifact, such as a piece of technology or even a physical organism. It is much harder to accept when it is so close to home, when it's our own deepest sense of self, our own internal universe that is revealed to be changing, moving, evolving.

Sleutels is a theorist who was partially inspired by Julian Jaynes, the author of one of the most popular books on consciousness written in the last decades, *The Origin of Consciousness in the Breakdown of the Bicameral Mind*. His radical thesis shocked readers in the 1970s by suggesting that consciousness, in the way we understand it today as an introspective interior space, only came into being relatively recently in human history, perhaps 2,500 to 3,000 years ago. He suggested that when the ancients claimed to hear voices of the gods, for example, this wasn't a quaint religious metaphor. They were actually hearing voices directing them to act. They were, in a sense, unable to internalize their own instinctual impulses within the context of a self-structure, and so they externalized them as outside forces. It was what we might call an extreme form of projection, to use current psychological lingo—although Jaynes attributed it to the developmental structures that had not yet formed in the brain. He noted that much of early literature represents a struggle to find a deeper sense of subjective selfhood, but for the most part the characters simply do not refer to themselves in any kind of way that suggests internal reflection.

> The characters of the *Iliad* do not sit down and think out what to do. They have no conscious minds such as we say we have, and certainly no introspections. [. . .] The beginnings of action are not in conscious plans, reasons, and motives; they are in the actions and speeches of gods.

Jaynes's thesis about changes in the brain remains highly speculative, but his basic idea, along with Sleutels's—that human consciousness has significantly changed, even during recorded history—resonates with the insights being revealed by a number of other evolutionary theorists. Indeed, if these insights are correct, then the very capacities of our awareness, the structures that make up the internal universe have undergone change, even dramatic change, over time. And some have taken this insight a step further, suggesting that not only has consciousness changed but that *it has evolved through a series of identifiable stages.*

To grasp the significance of this distinction, imagine for a moment how early geologists or paleontologists must have felt when they first discovered that the strata in an exposed rock face were not just random natural decorations but actually represented distinct eras in our planet's history—a pattern that could be found repeated independently in many different and far-flung locations. One layer of rock would reveal the geological and biological secrets of a particular epoch, while another layer revealed a very different world. Studying these layers of sediment, they could begin to understand how our planet has developed, and to recognize that the continuum of geological evolution has not been a steady curve but a series of distinguishable layers or stages. The same is true, we are now discovering, of the internal universe. It has evolved over time, and that change has not been random or erratic but has unfolded through a series of broadly identifiable stages.

What is a stage in the evolution of consciousness? When it comes to the internal universe, there are at least two ways of answering that question, because there are different ways to understand the evolution of consciousness. One is from the perspective of an individual and the other is from the perspective of the collective, and the tricky factor is that there is much overlap between the two. From the perspective of the collective, we can see, as we discussed in the last chapter, the way in which worldviews—those complex constellations of agreements and shared values—form and influence us in and through our shared, intersubjective space. In this sense, a stage would be a stable worldview. Unlike a geological period reflected in the strata of a rock face, a cultural stage is not simply a particular historical era, such as the Classical period or the European Enlightenment. The reason such eras stand out is because the core logic, the foundational values of new worldviews were birthed in those fertile times. But a particular worldview may endure through many historical eras, finding different surface expressions reflective of the society of the day while the underlying structural logic remains consistent. As philosopher David Owen writes, "One developmental-logical stage can underlie numerous societies with widely differing sociocultural elements." So a cultural stage of consciousness is not simply a stage of history. At the same time, history shows us that certain worldviews were undoubtedly dominant in particular eras and we can certainly track the evolution of dominant worldviews over the course of human history. One worldview coalesces as the primary outlook of a particular society, which then, for various reasons, breaks down and gives way to another in the long flow of history. The staged nature of the process is a generalization that only becomes apparent from a certain distance—lose yourself in the detail of any particular society or time period, and the general pattern dissolves. Attempt to draw perfectly clear lines between two worldviews and the effort will be in vain. But significant patterns exist for those willing to look through a larger historical lens.

The second way to think about stages in the evolution of consciousness is to consider the individual subjective psychological evolution over the course of a particular lifetime. We are accustomed to thinking about children this way—from toddler to teenager, we recognize that they go through temporary recognizable "stages" or "phases" that they (thankfully) outgrow. The field of developmental psychology maps levels not just of childhood but also of adult psychology that could also be referred to as important markers in the evolution of consciousness.

Arguably, the great challenge of the territory I am exploring in this chapter is recognizing that while these two dimensions of internal evolution—individual and collective—are distinct and have been mapped separately by most theorists, there is also tremendous overlap and interweaving between them. There is a profound relationship between the evolution of consciousness at an individual level and the evolution of culture at a social level, and we can find recognizable and even parallel stages in both.

In this chapter, I want to take a look at what we have come to understand about the stages through which human consciousness has evolved. I will begin with the philosopher Georg Hegel and the German idealists, because Hegel's evolutionary vision was so formative in shining a light on fundamental developmental logic, or the dynamics and dialectical process through which consciousness evolves. Then I will move into the realm of developmental psychology, in order to show how our increasingly sophisticated understanding of individual development has helped establish the reality that consciousness does evolve through stages, inspiring social theorists as they consider the broader canvas of cultural evolution. In exploring the more challenging territory of intersubjective development, I will turn to one of history's greatest guides to the vast landscapes of the internal universe—twentieth-century philosopher Jean Gebser.

These ideas are subtle, and they demand that we think in new ways about self and culture. As difficult as it was for the men and women of Darwin's time to wrap their minds around the idea that human beings had evolved their physical form and features over an extraordinarily long march of natural history, so too is it difficult for many in our own time to come to grips with the idea that the very nature of our consciousness, that inner self-sense that seems so fundamental to our humanness, has evolved through cultural history. But this notion has been developing, hand in hand with evolutionary philosophy, over the last two centuries.

Indeed, the idea that human consciousness has not only evolved but done so through an ascending series of worldviews is itself not new. Our intellectual history is rife with such systems of stages, many of which look quite crude and even silly from the vantage point of decades or centuries. Nineteenth-century intellectuals had many such systems, which unfortunately tended to relegate whole races and cultures to the dustbin of history, unworthy of respect or even simple tolerance. Our more recent embrace of plurality and egalitarian ideals has largely been a response to just such distasteful attempts to label certain people as evolved or advanced and others as primitive and barbaric. And unfortunately, mischaracterizations of the idea of evolution have played no small role in those deplorable schemes. With this in mind, we must be careful to bring tremendous nuance and subtlety to this inquiry, recognizing the real dangers of theorizing about cultural evolution. But even as we acknowledge these failings of history, the intellectual pendulum is swinging back and we are beginning to reexamine the idea that there may be legitimate differences between various worldviews or stages of cultural development that are critical to comprehend and dangerous to ignore. The idea of evolution has once again been a powerful spur, but now we are able to embrace that term in a way that far transcends the "survival of the fittest" ethos of the social Darwinists.

With the dawning recognition that both culture and psychology are caught up in an evolutionary movement, there has been a new enthusiasm that we might finally be able to crack the developmental code and understand more deeply the processes that have shaped and created human culture and human nature as we know them. Some have suggested that the worldviews that make up the internal universe play a role in cultural development that is not dissimilar to the function DNA serves in biological development. As we begin to understand the processes that go into the evolution of complex systems (and what is the human mind, if not a very complex system of processes?), we are recognizing that development isn't always just slow and steady. Gradualism may be the dominant inclination in scientific circles these days, but both developmental psychology and complexity theory suggests that evolution can also move in leaps and jumps, with periods of relative stasis mixed in with periods of rapid change. One organizing structure dominates until the system is pushed to its limits and then a rapid development takes place and a new organizing principle is formed—a new "dynamic equilibrium," as psychologist Robert Kegan describes it. We can see this process in psychological systems and even in biological systems. So stages are not exactly foreign to the dynamics of evolution; in fact, quite the opposite.

HEGEL'S EVOLUTIONARY PHILOSOPHY

"I wish to write a history not of wars, but of society," wrote the great French philosopher Voltaire. "My object is the history of the human mind, and not a mere detail of petty facts." Voltaire was one of the first historians to focus not on recounting dates and events but on trying to tell history from the inside out, to get inside the mind of history, and to understand the ideas and motives that made humans tick. Voltaire arguably represented the height of the European Enlightenment, and was certainly aware, in some rudimentary sense, of history as a developmental process progressing through stages, even though evolution was not yet an idea active in the culture at large. He wrote that he wished to track the "steps by which men passed from barbarism to civilization."

But if we are looking for the first truly robust evolutionary philosophy that set the groundwork for an understanding of stages, we need to move forward in time and northeast in direction—out of Voltaire's Paris, across the French and Belgian countryside, and over the Rhine, to a small German town on the edge of the river Saale: Jena. It was here, half a century before Darwin, that the great German idealists Georg Hegel, Friedrich Schelling, and Johann Fichte, inspired by the spirit of poet and philosopher Johann von Goethe, first explored how new ideas of change, development, and evolution might affect history and philosophy.* In this unlikely city, idealism, romanticism, natural science, Enlightenment philosophy, and religious inspiration all came together in a new evolutionary synthesis. As University of Chicago philosopher and historian of science Robert Richards has noted, Schelling was the first person to describe the concept of evolution as we understand it today. And as Napoleonic armies battled on the streets of Jena, Hegel completed his first major work, *The Phenomenology of Spirit*, in which he declared that "the Truth is not only the result [of philosophy] . . . the truth is the whole in the process of development."

Let us examine that last statement for a moment. Hegel is, for lack of a better word, *evolutionizing* the idea of philosophical truth. *Truth is the whole in the process of development.* He is liberating truth from the spell of solidity. Truth is not just found in this insight or that revelation; it is to be found in the very process of one idea giving way to another, and then being transcended by yet another, in the rough-and-tumble struggle of history. Truth is not static, he is saying, it is a process, a developmental unfolding. In order to appreciate any current philosophical idea, we need to understand its tributaries; we need to recognize the developmental process that has given it life. We need to take into account the give and take, the back and forth, the *dialectical* process, as Hegel calls it, as one stage of understanding gives way to another. "Hegel was the first to recognize that consciousness develops through a series of distinct stages," writes Steve McIntosh, "[and] among the first to understand that this process of development or 'becoming' is the central motif of the universe."

According to Hegel, any cultural or philosophical "truth" only becomes clear when seen in light of a larger framework of development. The worldview of any given era of history was "both a valid

---

* Though I refer here primarily to Hegel's work, I do so holding to scholar Fred Turner's observation that "ideas live less in the minds of individuals than in the interactions of communities." Fichte, Schelling, Hegel, and others formed the community of German idealists out of which these ideas sprang and though Hegel may have been the best synthesizer of the bunch, it is probably wrong to give undue credit to any one person.

truth unto itself and also an imperfect stage in the larger process of . . . truth's self-unfolding." Just as Theodore Dobzhansky would say over a century later that "nothing in biology makes sense except in light of evolution," in a sense Hegel was making the same point about philosophical ideas. They only come into true relief when seen in light of the larger historical evolution of ideas. "Truth is not a minted coin," he wrote, "that can be given and pocketed ready-made."

Hegel points out that those philosophers who reject a particular philosophical system in favor of their new and improved version usually fail to appreciate the mutual interdependence of even contradictory ideas—how one is built from the material of another in the flow of history, and all stand on a connected, growing synthetic meshwork of cultural propositions and truths. Indeed, in order to capture the way in which one developmental stage builds on the previous stage—preserving the core ideas of that previous stage while also transcending them and negating certain things as well—Hegel employs a German word that does not translate well into English. It is *aufheben*, and it means to both preserve and change. He writes:

> The bud disappears in the bursting-forth of the blossom, and one might say that the former is refuted by the latter; similarly, when the fruit appears, the blossom is shown up in its turn as a false manifestation of the plant. . . . These forms are not just distinguished from one another, they also supplant one another as mutually incompatible. Yet at the same time their fluid nature makes them moments of an organic unity in which they not only do not conflict, but in which each is as necessary as the other; and this mutual necessity alone constitutes the life of the whole.

Perhaps the closest translation in our own time comes from Ken Wilber, who uses the phrase "transcend and include" to explain the way in which one stage enfolds another in the march of evolution.

It's a principle that remains true for all kinds of organic and inorganic activities. It captures the way in which molecules transcend and include atoms, for example, which in turn transcend and include particles. Or one could say that one stage of culture is built upon yet moves beyond the essential insights of the stages that have come before. It is a universal principle of evolution that Hegel was beginning to glimpse in the nineteenth century, and it is essential for understanding the more sophisticated models of cultural evolution today.

Now, Hegel, Fichte, and Schelling have not always been loved by their philosophical brethren.* They were innovators, and their complex and obscure ideas did not always earn them acclaim from critics. Arthur Schopenhauer complained about all three of them and wrote that Hegel's philosophy was a "system of crazy nonsense" and would be remembered as a "monument to German stupidity." Karl Popper went so far as to accuse Hegel's philosophy of contributing to the rise of fascism in Germany.

It may well be true that for much of our contemporary intellectual world, the synthesis of the German idealists is more likely to be seen as the last gasp of a dying breed of philosophy. As philosopher Richard Tarnas writes, "With Hegel's decline there passed from the modern intellectual arena the last culturally powerful metaphysical system claiming the existence of a universal order accessible to human awareness." But with the rise of an evolutionary worldview, we can now more fully appreciate these extraordinary pioneers and their nascent but significant insights into the evolution of consciousness and culture. Hegel and his brethren may have been the last of the old, but they were also the first of the new. They helped acclimate us to the idea that consciousness evolves and that there are discernible patterns in its historical development. And while that idea would certainly be abused again and again by people looking to situate themselves as the pinnacle of their own evolutionary schema, the seed idea remains essential. We can now appreciate the insights of the idealists as the necessary beta versions of the evolutionary principles that today are maturing into more pragmatic tools with which to understand everything from politics to ecology to religion.

"THE LAWS OF THOUGHT ARE THE LAWS OF THINGS"

In his latest book, *Evolution's Purpose*, Steve McIntosh identifies developmental psychology as "the branch of social science that deals most directly with the evolution of consciousness." Indeed, this rich tradition, which started out studying the processes and stages through which children develop and later expanded to include adult development as well, is critical for understanding the kinds of research, thinking, and perspectives that have given birth to our new appreciation of how the internal universe evolves. In fact, after Hegel and the idealists, it was the pioneers of this then-brand-new field who took up the idea of stages in consciousness, giving the notion some much-needed theoretical backing and empirical weight.

The true legend in this field is the Swiss developmental psychologist Jean Piaget. If you start up a conversation about stages of human development, sooner or later someone is going to bring up his name. His breakthrough work with children in creating a "staged" model of cognitive development had a huge effect on culture. He changed forever the way we understand children and, by extension, adults as well. But if you ask historians about the pioneers of developmental psychology, another figure comes up—the unsung early hero of the field. And his name is much more directly associated with evolution: James Mark Baldwin.

Baldwin was an early twentieth-century genius who is probably known best for a process called the Baldwin effect or Baldwinian evolution. Still considered seriously today, it suggests that it is possible for individuals of a species to alter their own genome through a sustained change in behavior—to turn a learned habit into an instinctual one. The idea is that some behavioral changes, if continued over multiple generations, can eventually become inherited, and actually written into the genetic code. What starts as habit can eventually become instinct. Lactose tolerance is often used as an example—initially it was common only to children, but as humans became pastoral and started domesticating cows, eventually childhood lactose tolerance extended into adulthood. Darwinian selection pressures begin to favor lactose tolerance.

Baldwinian evolution has gone in and out of favor over the years, but particularly with the emergence of epigenetics in recent decades, science has come to appreciate a whole new level of plasticity in the expression of the genetic code. As a result, Baldwin's original propositions have recently taken on even more scientific relevance, and his name increasingly turns up in current evolutionary literature. He was perhaps the first psychologist for whom, in the words of author Henry Plotkin, "evolution was an absolute conceptual anchor." Indeed, to read Baldwin today is to appreciate just how far ahead of his time he truly was.

Just as Hegel exploded the idea of a fixed, static notion of truth, Baldwin exploded the idea of a fixed static notion of mind. Through observing the development of his own daughter, he saw that her cognitive structures were developing through specific stages over the course of her childhood. Baldwin's careful observation of this process and his work to eventually build an entire theory of cognitive development around those observations was one of the great breakthroughs in psychology of that time. Many people today don't even realize that only a couple hundred years ago we used to treat young children as if they were little adults, with less knowledge but essentially the same capacities. Our contemporary understanding that children go through a developmental process is in part due to Baldwin's unique genius combined with his deep appreciation for the power of an evolutionary perspective.

Called a spiritualist metaphysician by one historian, Baldwin and his interests straddled many worlds, and his work has been described as a bridge between "social and cognitive theories of development," bringing to mind later theories such as Spiral Dynamics (which we will explore in the next chapter) and the philosophy of Jürgen Habermas, which cross the lines between those two realms. Baldwin also wanted to break down the separation between mind and nature. "The laws of thought are the laws of things," he once wrote, and in fact, he may have been one of the first to try to bring psychological development and biological development under the same general umbrella, anticipating the cross-disciplinary nature of an evolutionary worldview in statements like: "General biology is today mainly a theory of evolution, and its handmaid is a theory of individual development."

Some have even suggested that if it weren't for a little incident at a brothel in Baltimore (a "colored" brothel, God forbid) that got Baldwin kicked out of Johns Hopkins University and sent him into exile from America, his reputation today might rival Piaget's in terms of developmental psychology. As it turned out, his breakthrough proposition that human beings evolve through a very specific series of developmental stages—which he called pre-logical, quasi-logical, logical, super-logical, hyper-logical, and even mystical stages of consciousness—remained relatively obscure, although evidence suggests that it did help to inspire Piaget in his formative years.

Later, in the hands of developmental theorists such as Lawrence Kohlberg and Carol Gilligan, who studied moral development, and Jane Loevinger, who mapped ego development, Baldwin's essential multistage evolutionary structure would grow and expand into more sophisticated and empirically studied evolutionary models. And with those theorists would come a deeper understanding of just how evolution actually works in the interior of the self and what exactly this thing called a stage of development is anyway. But in the early twentieth century, even as Piaget went on to introduce the world to a new stage-centered psychology that became enormously influential, Baldwin's name faded from the limelight. And with him went the critical realization that there is more than a little connection between evolution as a scientific idea, a philosophical notion, and a psychological template.

Today in academia, if there is anyone who is a successor to Baldwin, it would have to be Harvard psychologist Robert Kegan. His work on stages of psychological development has been groundbreaking, and his books, *The Evolving Self* and *In Over Our Heads*, are read by college students around the country. Kegan's understanding of stages has made a seminal contribution to the field and has built on the work of Baldwin and Piaget before him. Suffice it to say that when we look at the development of this field from Baldwin to Piaget to Kegan, we see an emerging recognition that human consciousness is plastic and malleable in a way that continues to be underappreciated even today. Developmental psychology has broken the spell of solidity in multiple respects. First, it helped us understand that children develop, that they do not possess pre-given adult minds but go through psychological stages of growth on their journey into adulthood. And now we are beginning to understand that adults also can develop in remarkable ways. As Kegan explains, "The great glory within my own field in the last twenty-five years has been the recognition that there are these qualitatively more complex psychological, mental, and spiritual landscapes that await us and that we are called to after the first twenty years of life."

So what is the relationship between these individual stages and cultural worldviews? As we learned in previous chapters, we can't think of worldviews as purely individual matters, or even as an amalgam of individual minds. They are also cultural organisms. They are constellations of values that may be created by humans but that also have their own systemic existence. So are the worldviews that have characterized the development of human culture over millennia similar to the psychological structures through which the self develops? That's the $64,000 question. Is there an individual/ cultural parallel to the old biological idea that "ontogeny recapitulates phylogeny"? That was German evolutionary theorist Ernst Haeckel's famous phrase, describing the way in which the developing fetus can seem to go through stages that resemble the evolutionary development that life went through in the journey from cells to fish to mammals.* The idea is that the development of the individual recapitulates the development of the species. Haeckel's proposition has been disproved as any kind of absolute law, but there is nevertheless an acknowledged correlation. In the same way, there has been much speculation that the development of the individual mind recapitulates the evolution of culture. Do Piaget's stages of psychological development correspond to the evolution of cultural worldviews over the course of history? Again, we would be wise to steer clear of any overly facile or determinative correlations. But important parallels exist nonetheless. As McIntosh puts it, "Even though individual development and cultural evolution are not identical, the developing mind does reveal patterns in its unfolding, and these patterns resonate with the historical unfolding of culture that occurs on an evolutionary time scale."

German philosopher Jürgen Habermas is one of the few developmental theorists to take up the challenge of bridging these two domains. He argues that both psychological development and cultural development are ultimately based on linguistic structures, and it has since been convincingly demonstrated by Piaget, Kohlberg, Kegan and others that if a developmental logic exists in the maturation of the individual ego, then there is every reason to suspect that social evolution would exhibit a similar trajectory. The developmental logic that informs one would naturally inform the other, because the same underlying structures are responsible for both.

---

* For more on Haeckel, see *The Tragic Sense of Life* by Robert Richards (Chicago: University of Chicago Press, 2008).

Habermas also makes the critical distinction we touched upon in the last chapter, between the deep structures that define a stage in consciousness and the more superficial surface manifestations of that stage that might vary greatly from person to person, from society to society. This helps answer one of the common concerns about a stage-oriented view of cultural development—that it does not account for the significant differences that exist among unique cultures around the globe. For example, how could stages that we might see in the historical development of North America and Europe really be applicable in an Asian context? The answer is that two cultures can express significantly different forms and features of surface content contingent upon a host of unique historical circumstances and still be constructed on the same underlying core worldview or deep structure of consciousness. As China and India embrace native forms of modernism, for example, we have the opportunity to observe this principle at work, to see the ways in which each nation brings its own unique character to the leap it is taking, and yet at the same time, both exhibit patterns that are more than reminiscent of the process Europe and North America went through in the last couple of centuries.

More research will be needed to establish the relationship between the evolution of individual consciousness and the evolution of cultural worldviews. But however connected these dimensions may be, there is one big difference when it comes to studying them. It's much easier to look at the stages of an individual mind in the space of a lifetime than it is to look at the cultural stages that emerged in the dim reaches of history. Where are we to look for historical evidence of the development of a dimension that is invisible to the human eye and not easily discernable by the instruments of science? Evolutionaries who choose to study this aspect of life's unfolding face a challenge. "Unlike paleontology, where the outlines of ancient life-forms are open to study," writes scholar of mysticism Gary Lachman, "in trying to understand what a consciousness prior to our own might have been like we run into the fact that there are no 'fossil remains' of previous consciousnesses available for our inspection." We must improvise, he suggests, by studying carefully the outward expressions of human culture, which naturally reveal something about the character and nature of human consciousness. In the artifacts of culture, he writes, we can see "the imprint of human imagination . . . the mind pressing itself upon the material world." We must look at these "mindprints" in order to catch a glimpse of the consciousness that created them.

Perhaps the best such archeologist of the mind in the last century was philosopher Jean Gebser. Even while living a tumultuous early life that saw him narrowly escape both Franco's fascists and Hitler's Nazis, he was able to get inside history, so to speak, to peer into the internal universe and start to map the development of structures of consciousness and culture from the inside out.

"A SUMMONS TO CONSCIOUSNESS"

If James Mark Baldwin, as University of Chicago historian Robert Richards suggests, was a spiritually alive person who "felt the beat of consciousness" in his own life, then Gebser must have felt the music of the entire orchestra. He was, as William Irwin Thompson described him, a "brilliantly intuitive intellectual mystic" whose novel theory of how human culture has evolved through four distinct "structures of consciousness" was born out of a powerful spiritual epiphany. In 1931, Gebser, a struggling German thinker living in exile from his native land in Spain, underwent a mystical awakening that convinced him that a new form of consciousness was beginning to appear in the West. He called this consciousness "integral" and distinguished it from the other forms of consciousness that had come before—which he labeled *archaic, magical, mythical, and mental-rational.* Over the next two decades, this singular insight would inform all of his work and he would tirelessly explore the implications of his integral awakening, eventually producing his masterpiece of philosophy, mysticism, and history entitled *The Ever-Present Origin*. For Gebser, integral consciousness is a form of consciousness that is concerned with "integrality and ultimately with the whole." He distinguished this integral consciousness from the mental-rational stage that had come of age in the Renaissance with the "discovery of perspective" and of three-dimensional space. A friend of Picasso, Gebser was quite concerned with perspective, and often used art as a window into the consciousness of any given historical period. He felt that the artistic discovery of perspective in space was linked to the "entire intellectual attitude of the modern epoch," and was essential for the scientific-technological world we inhabit today. For Gebser, the dawning of integral consciousness would serve to reintegrate the mental-rational world that had grown so fragmented, and also allow previous stages of consciousness to resurface in our awareness. They would become "present to our awareness in their respective degrees of consciousness."

Surprisingly, Gebser was not a fan of evolution. He referred to his "structures of consciousness" not as evolutionary stages but as mutations. He saw evolution through the narrow window of his own time, as a limited, materialistic conception of the dominant mental-rational worldview. And so he rejected the term. It is only in retrospect that we can see his insights as being evolutionary in the richest and broadest sense of the word.

Reading Gebser is like reading an impressionist painting; the effect is not in the details of the argument built step by careful step but in the sudden flashes of revelation that one receives as Gebser paints an evocative portrait of how consciousness evolved through the millennia. Unlike many other thinkers, Gebser is not enthralled by the Flintstones fallacy. Indeed, he gives us a peek inside structures of consciousness that were active hundreds and thousands of years ago, bringing forth insights that startle us with their strangeness while simultaneously stirring memories and familiar glimpses. It is worth noting that while Gebser acknowledges that many of these stages remain present in consciousness and culture today, his particular gift is bringing alive a sense of the original versions of these earlier stages as they would have been experienced when they first emerged back in the far reaches of time, unmediated by the overlay of subsequent evolution. It has been left to other theorists, as we shall see in the next chapter, to shine light on the contemporary expressions of these structures and help us figure out how to navigate them.

The earliest structure in Gebser's model is called *archaic*. It is ancient and hard to analyze through historical data, as any reference to it is already coming from a perspective that has begun to move beyond it. "Written evidence is itself indicative of a transitional period," he writes. Archaic man was indistinguishable from the world and the universe, Gebser suggests, living in a state of consciousness in which there is simply no differentiation from nature. He notes only a few sources of information about these early humans, notably Chang Tzu's statement about the ancients in the fourth century BC: "Dreamlessly the true men of earlier times slept." The dreamlessness suggests the dormancy of a certain level of conscious awareness.

The next structure is *magic*. Gebser acknowledges that these stages are fluid among different cultures and time periods, and there are not clear lines between one and the other, only transitional times. The distinguishing feature of the magic stage is that humans have now been released from their "harmony or identity with the whole." This is a time of human fighting to step outside of nature, to free oneself of the inertia of nature, to be independent of nature. "He tries to exorcise [nature], to guide her," Gebser writes. "He strives to be independent of her, then he begins to be conscious of his own *will*."

This is a time of impulse and instinct, vital forces and group ego, of ritual and rainmaking, totems and spell casting, miracles and taboos, and the barest beginnings of individuation. The more that magic man is able to free himself from that "sleep-like consciousness," the more he becomes an individual, a unity, as Gebser puts it. At this point he is unable to "recognize the world as a whole, but only the details which reach his sleep-like consciousness and in turn stand for the whole. Hence the magic world is also a world . . . in which the part can and does stand for the whole." This association between the whole and part allows for a world in which everything "intertwines and is interchangeable"—a world of symbol, a world in which the picture of the bison and the bison can be interfused, just as poking a voodoo doll can elicit a reaction in the person represented, or a ritual sacrifice can elicit real but vicarious suffering.

As always, Gebser uses art to illustrate, demonstrate, and support his primary descriptions. He notes, for example, that art illustrating this structure has several interesting features. First, there is an interesting "mouthlessness" in early sculptures and works of art from this period. Magic humans, or at least their early expressions, had little need of language, as the group ego and the general egolessness of individuals allowed for a certain telepathic communication and submission to the vital needs of the clan. Sounds are more important in this structure. Language is still in its early stages. "Only when myth appears does the mouth, to utter it, appear," writes Gebser, and he draws parallels between this and preverbal stages of development in infancy. He also remarks on the way in which early art illustrates the overriding fusion between humans and the natural world. Some paintings even seem to have humans and nature essentially merged in this "spaceless, timeless" world, a point that he shows again and again in images representative of the time period. Here again, we see the magic human's primordial embeddedness in nature and the struggle to be free.

This release from nature is the struggle which underpins every significant will power-drive, and, in a very exact sense, every tragic drive for power. This enables magic man to stand out against the superior power of nature, so that he can escape the binding force of his merger with nature. Therewith he accomplishes that further leap into consciousness which is the real theme of mankind's mutations.

This remarkable and deeply inveterate impulse to be free from miracles, taboos, forbidden names, which, if we think back on the archaic period, represents in the magic a falling away from the once-prevailing totality: this urge to freedom and *the constant need to be against something* resulting from it (because only this "being against" creates separation, and with it, possibilities of consciousness) may be the answering reaction of man, set adrift on earth, to the power of the earth. It may be curse, blessing, or mission. In any case, it may mean: whoever wishes to prevail over the earth must liberate himself from its power.

Next in Gebser's schema is the *mythic* stage. Again, transitional phases are acknowledged, and there are inevitably differences in terms of timing across different cultures and geographical regions. Human cultural evolution is not some kind of monolithic movement forward, everyone together in lockstep. This is an important point and often misunderstood by critics. Moreover, Gebser suggests that there are both positive and negative expressions of each of these stages, time periods in which they are rising with growth, vitality, and all the energy of new emergence and time periods in which they are fading, losing their vitality, and entering a more "deficient" period.

With the arrival of mythic human culture, we have also the beginning of history, literally. The magical is prehistoric; it lies "before our consciousness of time," as Gebser puts it. Recorded history is itself a sign of the emergence of a new consciousness, a sign that the timeless consciousness of the magic culture is giving way to a new sense of time and history. Time is beginning to come into human awareness, an emerging sense of temporality, which will lead into new stories of human origins and mythical cosmologies. And with these new myths and cosmologies, there is a new sense of language and words to express them, a new oral tradition that places emphasis on the power of the "vocalized mythical narrative" ("In the beginning there was the Word"). Words take on great significance in the mythical structure as conveyors of psychically rich meaning and vitally infused content. Even today, Gebser notes that the mythic structure exists as a pictorially rich, imaginatory world, characterized by the imagistic nature of myth, which alternates between "magical timelessness and the dawning awareness of cosmic periodicity."

Gebser, as I mentioned, was fascinated by artistic perspective and notes that mythic humans evolved a new sense of perspective, an "unperspectival" two-dimensional polarity signified by the circle, a cyclical mindset that "encompasses, balances, and ties together all polarities as the year, in the course of its perpetual cycle of summer and winter, turns back upon itself; as the course of the sun encloses midday and midnight, daylight and darkness, as the orbits of the planets, in their rising and setting, encompasses visible as well as invisible paths and returns unto itself." And this polarity is captured not just in physical expression but in psychic and religious realities as well. For example, in the invocation of the "sub terrestrial Hades and . . . the super terrestrial Olympus." Later, of course, we see this same polarity in the belief in a heaven above and hell down below.

The emergence of the mythic represents a new cognizance of the internal universe, the world of the soul, as Gebser explains it. Awareness is increasing; the inner world is becoming larger, richer, and more conscious. One sign of this increasing consciousness, he notes, is the inclusion of human anger and wrath in early literature, a demonstration of a new individuation and self-assertion that marks the human ego striving to break free of the constrictive bonds of the group, the collective, the clan. We see this in the wrath of Moses, "awakener of the nation of Israel," and in the hero mythologies like the *Iliad*—perhaps the earliest work of Western literature—which actually begins with the statement: "Sing, O goddess, the anger of Achilles, son of Peleus." In this light, we can see how this mythical shout-out to Achilles is simultaneously a tribute to the awakening of a self-directed sense of self, or as Gebser puts it, a "summons to consciousness."

The *mental* structure of consciousness is more familiar to us, as it still represents the dominant stage of our culture today. I have tended to refer to it as *modernist* in this book, the phase of consciousness or worldview that came of age in the culture at large during the European Enlightenment. Gebser places its true beginning in the Axial Age, however, in Hellenic Greece, the fountainhead of the Western mind. It is the emergence of the rational mind of directed and discursive thought, another step in the long journey of the individuated ego to free itself from the domination of nature, the vital bonds of the clan, and the psychic energies of myth. Gebser, ever the subtle mythologist, cites not only the philosophers of Athens as the birthplace of the mental-rational, but also the birth of Athena, the goddess of wisdom who is born from the cloven head of Zeus. This goddess of "bright . . . clear-thinking" is an owl-eyed deity who can see into the unconscious, unawakened darkness, is the namesake of the Athenian mind, and is destined to become a patroness of the sciences. It is here, as Gebser notes, speaking from the perspective of the European mind in the middle of two disastrous world wars, that the mental-rational "world came into being, our world which is now perhaps coming to a close. Anyone who perceives the end and knows of its agonies should know of the beginning."

Gebser notes that thinking or "thought forms" in the manner that we think of mentality today is really a feature of this new mental-rational structure. We cannot really speak of thought as we know it in the original mythical structure, he suggests. Thought forms, in that earlier structure of consciousness, are different, more like "being-thoughts"—a designation that illustrates the more embedded character of individual consciousness in the mythical structure. In the mental stage, there is an unambiguous "I" doing the thinking ("I think, therefore I am"), but in the mythic world, humans could not so easily distinguish their conscious agency from the thought forms that occupied their psyches. "Whereas mythical thinking, to the extent that it can be called thinking," he writes, "was a shaping or designing of images in the imagination, discursive thought is fundamentally different. It is no longer polar related, enclosed in . . . but rather directed toward objects . . . and drawing energy from the individual ego." And in the magic structure we lose even more dimensionality; Gebser calls thought forms here "being-in-thought"—an expression of a less individuated ego, more immersion; a world not of discursive, representational thought or psychic experiences but of vital experiences in which the individual being is almost completely subsumed.

The *mental* is a structure dominated by men, mentality, matter, and materialism in which we move from mythology to philosophy, in which "man is the measure of all things" and man measures all things. The emergence of this new stage is earthshaking in that "it bursts man's protective psychic circle and congruity with the psycho-naturalistic-cosmic-temporal world of polarity and enclosure. . . . Man steps out of the two-dimensional surface into space, which he will attempt to master by his thinking. This is an unprecedented event, an event that fundamentally alters the world."

More than even his unique and, in most respects, unprecedented description of these stages of consciousness, Gebser also captures the agonies and the ecstasies of the heroic struggle of human evolution, the hard-won victories of our increasing consciousness, and the increasing freedom and dimensionality of our experience. For him, it is unquestionably a spiritual process, and our reflection on it a chance to perceive the very "seedlings of the future" that will light the way to the next emergence. And it reveals both the wonderful and terrible consequences of our complex journey from the unity of a primordial fusion with nature into the freedom, light, and power of a conceptual, rational universe.

> And it would be well for us to be mindful of one actuality: although the wound in the head of Zeus healed, it was once a wound. Every "novel" thought will tear open wounds . . . everyone who is intent upon surviving—not only earth but also life—with worth and dignity, and living rather than passively accepting life, must sooner or later pass through the agonies of emergent consciousness.

Through Gebser's evocative writing, we can begin to see human evolution through this rich tapestry of stages and structures. What happens when worldviews collide? When we see historical events through the lens of the evolution of consciousness, it changes our perspective in important ways. We become less focused on the material manifestations of the worldview and more on the inner dimensions, on the internal structures and where they are situated in evolutionary time. For example, in one particularly evocative passage that illustrates this perspective, Gebser describes the reasons why the Aztec civilization of Mexico gave way so easily to the Spanish conquistadors in the sixteenth century. How could so few conquer so many, so easily? He begins by referring to an ancient Aztec account of how Montezuma tried to send his sorcerers to cast spells on the Spanish. Their failure was not just a failure of will or effort or technology or manpower or weaponry but of consciousness, he explains. The Aztec culture was reflective of Gebser's magical and mythical stages, whereas the Spanish were more rooted in the mental-rational. Montezuma sent his spell casters, priests, and soothsayers to intercept the Spanish, but as the manuscript recounts, they "could not reach their intent with the Spanish; they simply failed to arrive." Current theorists may attribute this to a superstition on the part of the Aztecs, but Gebser suggests that "authentic spellcasting, a fundamental element of the collective consciousness for the Mexicans, is effective only for the members attuned to the group consciousness. It simply bypasses those who are not bound to, or sympathetic toward, the group. The Spaniards' superiority, which compelled the Mexicans to surrender almost without a struggle, resulted primarily from their consciousness of individuality, not from their superior weaponry. Had it been possible for the Mexicans to step out of their egoless attitude, the Spanish victory would have been less certain and assuredly more difficult."

Gebser's account of this encounter between cultures gives a hint of what it means to begin to look at history through the lens of the interior evolution of consciousness rather than exclusively focusing on the more material world of technology, economics, science, or politics.

In addition to delineating these stages of history, Gebser was also tracking the awakening of a new worldview—one that he had glimpsed as a young man, and labeled "integral." But the new is also about the old, in the sense that the dawning of integral consciousness includes reawakening both our historical knowledge and inner connection with previous structures, making them more conscious, more transparent, more accessible. But this is neither a return nor a regression. We cannot go back; the way forward is greater consciousness, not less. We must not try to reinhabit or romanticize cultural structures inappropriate for our own time. We see this happening, for example, in postmodern society's near obsession with indigenous cultures' wisdom and connection to nature, which was really, as Gebser shows us, an *immersion* in nature. And we see it in those who retreat into mythic religious worldviews, seeking the established security of accepted meaning and a sense of place.

Once again, I should point out that while Gebser readily admits that we may see these early stages active in our own psyches and that they may be prominent and even dominant in certain cultures today, their original historical character would have differed significantly from their expression in our own time. A mythic-stage culture, when it represented the very leading edge of human evolution, valiantly struggling to liberate itself from immersion in the magic stage, would inevitably be worlds apart from a mythic worldview existing in the context of a postmodern globalizing world, much of which has left mythic consciousness far behind in the misty reaches of the past. Each of these structures continues to mutate and evolve, some quite significantly, even if certain core elements, the deeper structures of consciousness that underlie each stage, remain roughly the same. This distinction between the deep underlying structures of these stages of consciousness and their more relative, changing, contingent surface expressions is a critical feature of a robust view of cultural evolution. More important, it illuminates the world around us. For example, it explains why we can have reactionary Islamic fundamentalism that is both a very contemporary response to modernity and at the same time expresses core elements of a very ancient mythical structure of consciousness.

Gebser would continue, throughout his life, to look for examples of the new *integral* consciousness emerging in world culture, eventually traveling to India, where a second spiritual illumination occurred at Sarnath, the site of the Buddha's first teaching, which he described as a "transfiguration and irradiation of the indescribable, unearthly, transparent 'Light.'" In the introduction to the second edition of *The Ever Present Origin*, published in 1966, he noted in a passage that burnishes his credentials as a significant Evolutionary that among examples of this new stage of consciousness, "the writings of Sri Aurobindo and Teilhard de Chardin are preeminent."

Gebser's rich and extraordinary insights were inspired by his own awakening glimpses of integral consciousness. And his work

has done a tremendous amount to contribute to the body of philosophical thought that is emerging today around the notion of stages
of cultural development. He also set the stage for a whole generation
of thinkers to research, fill out, and refine the basic patterns he had
identified and begin to work with them as they appear in culture
today. As we will see in the following chapter, understanding the
consciousness of yesteryear is essential not only for appreciating the
fact that consciousness *has* evolved but also for enabling it *to* evolve,
in our very own time, even in some of the most volatile places on our
ever-smaller planet.

CHAPTER TEN

Spiral Dynamics: The Invisible Scaffolding of Culture

A developing brain is a sort of snowballing cognitive leviathan that adapts to everything and anything close to it. Learning is one aspect of extreme plasticity, and creativity another. Any species that can do such things as play with the world, imagine it, remember it, and expand its circles of experience . . . will ultimately start to experiment with its own fate.

—*Merlin Donald*, A Mind So Rare

In my travels around the progressive spiritual and philosophical world over the last two decades, I have met many unique, contradictory, endearing, and surprising characters, but none quite prepared me for my meeting with Don Beck—a tough-talking Texan academic activist with a unique perspective on cultural evolution. With his soft drawl and his mixture of brashness and charm, Beck helped deepen my appreciation of that powerful insight we have explored in the last chapter: that consciousness and culture evolve through identifiable stages and structures. It's a bold and controversial proposition, but it's one that is definitely worth the time and investment to understand. And in Beck's hands, this insight takes on a particular cultural relevance. Whereas much of Jean Gebser's work was concerned with envisioning forms of consciousness as they first emerged in our cultural past, Beck is concerned with those stages as they continue to manifest today. As I've mentioned, the sequence of worldviews that define the trajectory of culture's unfolding are not simply features of our history—they still exist as stable organizing systems for societies around the world. Understanding the reality and nature of these worldviews is one of those ideas, as they say, whose time has come, and I suspect it will play a critical role in making sure that human beings do not repeat the mistakes of the nineteenth and twentieth centuries in the twenty-first and twenty-second.

While Beck's system incorporates some of the basic ideas of Gebser, Hegel, and developmental psychology, Spiral Dynamics, as it is known, is a more practical and pragmatic way to look at the evolution of worldviews. It is the brainchild of maverick psychologist Clare Graves, who was Beck's friend and mentor before his death in 1986. The basic idea of Spiral Dynamics is quite simple—deceptively so. There are eight stages or "value systems" or worldviews (Beck currently refers to them as "codes") that form the basic structures of human psychology and sociology. These stages make up an ascending evolutionary spiral that both individuals and cultures will pass through as they develop—psychologically, socially, morally, spiritually. Beck refers to these as "bio-psycho-social-spiritual" systems that form a sort of invisible scaffolding in our consciousness, unseen but influential cognitive structures that condition our perspectives and our values analogous to the way DNA influences but does not exactly determine the forms and features of an organism. Indeed, just as Abraham Maslow, the mid-twentieth-century pioneering psychologist, was tracking a hierarchy of needs, Spiral Dynamics tracks a hierarchy of values. In fact, the relationship between those two developmental systems goes beyond mere systemic resemblance; Maslow and Graves themselves were friends and colleagues.

Beck and I first met in 2002, when he visited the offices of *EnlightenNext* in Massachusetts. He was in his seventies and I was in my thirties, but as luck would have it, we had a couple of things in common more important than age—a passion for the dynamics of evolution and a love for the sport of American football.

I grew up in Oklahoma, so I know something about that unique species of American male known as Texans. First, they tend to have a chip on their shoulder and an independent streak. Beck has both in spades. And second, they love football. So during those first encounters with Beck and Spiral Dynamics, my colleagues and I would spend hours and hours discussing the ins and outs of evolutionary stages with Dr. Beck, and then he and I would slip away, find a television, and watch college football.

Now, as my British wife will attest, when I watch football, especially University of Oklahoma football, I undergo a rather startling personality change. Temporarily, I leave behind my mild-mannered exterior and a whole subpersonality comes to the forefront of my consciousness. It's as if I'm getting in touch with my tribal roots, with warriorlike values of power, will, and domination that are not so prominent in my everyday personality. A whole new attitude emerges in my consciousness, which I suspect is more related to ancient tribal wars than anything I'm engaged with currently. It is also a predilection that runs in the family (as well as in the state). When my wife first met one of my cousins, who still lives in Oklahoma, my cousin congratulated us on our recent marriage and then quickly asked my wife with some concern, "Have you seen him watch football yet?"

Thankfully she has, and we are still happily married, but the larger point is that my temporary change of character speaks to the theory of Spiral Dynamics. Spiral Dynamics suggests that, as Gebser also believed, each of the major value systems represents an internal structure that exists within each of us. These can be reactivated at any time, depending on the circumstances of our lives. I'm watching football, and for a couple of hours I can experience, in some rudimentary way, the values and emotions more closely associated with a "might makes right" world of Attila the Hun than with a modern democracy. Now, that doesn't mean that I lose all control and turn into a tribal warrior, but it does mean that given the right conditions, any of us can, to greater or lesser degrees, reinvoke or reinhabit perspectives and attitudes whose most salient features were formed in earlier eras. Just as evolutionary psychology makes a powerful argument that many of the habits, traits, and impulses that make up our modern character were originally formed deep in our evolutionary past, Spiral Dynamics argues that many of our personal values are actually quite impersonal, formed in the evolutionary cauldron of human history.

"People are trapped in history and history is trapped in them," wrote the great American author James Baldwin. Baldwin was talking about race, but the statement also captures the way in which the evolutionary history of the species is unavoidably reflected in the interior of our individual psychology. Indeed, according to Spiral Dynamics, we are not blank pages on which we may write any drama we please. No, we are living *in* the developmental drama of history, and the sooner we recognize the true contours of that script, the more influence we can have on how the play unfolds. In that sense, "trapped" is the wrong word, but we are living in history and history is living in us.

As discussed in the previous chapter, one of the challenges for developmental theorists is understanding the relationship between individual and collective, between cultural worldviews and psychological stages of development. Spiral Dynamics is interesting in that it does seem, at least to some degree, to apply to both. But adherents have also been criticized for blurring the distinctions and drawing unproven correlations. In the pages that follow, I will no doubt be guilty of this myself, but I do so consciously for the sake of illuminating the genuinely powerful features of this perspective, and I ask readers to hold these distinctions lightly.

Beck himself was a professor of sociology at the University of North Texas when he came across Graves's work in 1974, in an article in *The Futurist* magazine entitled "Human Nature Prepares for Momentous Leap." He was immediately taken with the ideas contained in the essay. The open-ended evolutionary nature of Graves's theory struck him—the feeling that human nature was not some fixed event waiting to be mapped and understood but an unfinished, malleable, evolving system that was still in process, still adapting, still changing. In fact, we might say that he sensed in Graves's work a shattering of the spell of solidity in relationship to human culture. "The error which most people make when they think about human values is that they assume the nature of man is fixed and there is a single set of human values by which he should live," Graves declared right at the beginning of the 1974 article. "Such an assumption does not fit with my research. My data indicate that man's nature is an open, constantly evolving system, a system which proceeds by quantum jumps from one steady state system to the next through a hierarchy of ordered systems." Never one to contemplate people and big ideas from afar, Beck was soon on a plane to upstate New York, where he met the man behind this fascinating new model. They hit it off at once (Graves loved sports too) and spent hours together discussing the meaning of this new theory.

The trip was nothing short of revelatory. When Beck got back to Texas, he completely changed his research direction to further explore the implications of Graves's theoretical model. But he also took a new interest in other related models of psychological and moral development that had been popping up in the decades since World War II. So he put his Texas-sized cowboy boots on the ground and headed out to meet the great developmental theoreticians of the day, such as legendary ego psychologist Jane Loevinger, and Lawrence Kohlberg, the celebrated Harvard theorist of moral development. But despite these illuminating visits, he found no work that had the depth of Graves's theory, so he stayed in close touch with his new mentor, developing a friendship that would span the rest of the older man's life and define the younger man's career.

Graves was a maverick whose ideas ran counter to the dominant theories of behaviorism of the day, and even to the more progressive direction of humanistic psychology. He attributed some of the originality of his thinking, his cross-disciplinary comfort level, and "his ability to see differences" to the unusual diversity of perspectives he encountered during his years at Western Reserve University. This formative experience may have been part of Graves's intellectual salvation, perhaps forging that generalist perspective characteristic of so many Evolutionaries. But he never achieved the reputation or had the influence of other significant theorists. And were it not for the efforts of Beck and another important colleague, Chris Cowan, who worked with Beck to mold Graves's ideas into a more contemporary form, Graves's unusually integrated approach to human development might still remain obscure. In the light of history, we might say that Graves's rejection by his peers was simply a result of being far ahead of his time. Yet it also points to one of the most salient aspects of his theory—that "life conditions" play a critical role in development. As times change and culture evolves, the kind of people who take the cultural center stage change as well. In war, we need generals and men willing to fight. In peacetime, we celebrate very different kinds of achievements. Similarly, as culture evolves, we newly appreciate those theorists and theories whose contributions correspond more to the needs of our own moment than the time when they were alive. Today, an evolutionary theory like Spiral Dynamics is filled to the brim with new forms of explanatory power that are more suited to helping us sort through the culture wars and the so-called clash of civilizations. And so we might say (using Darwinian language) that it has become more suited and "fit" for the cultural environment of the twenty-first century.

UNFOLDING, EMERGENT, OSCILLATING, SPIRALING BIO-PSYCHO-SOCIAL SYSTEMS

"Briefly, what I am proposing," Graves writes in the article that first introduced Beck to his work, "is that the psychology of the mature human being is an unfolding, emergent, oscillating, spiraling process marked by progressive subordination of older, lower-order behavior systems to newer, higher-order systems as man's existential problems change."

The Spiral of Cultural Development*

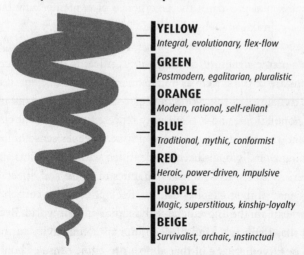

YELLOW
Integral, evolutionary, flex-flow

GREEN
Postmodern, egalitarian, pluralistic

ORANGE
Modern, rational, self-reliant

BLUE
Traditional, mythic, conformist

RED
Heroic, power-driven, impulsive

PURPLE
Magic, superstitious, kinship-loyalty

BEIGE
Survivalist, archaic, instinctual

*Colors refer to the Spiral Dynamics model, based on the work of Clare Graves, Don Beck, and Christopher Cowan

By "older, lower-order" and "newer, higher-order" systems, Graves was referring to the systems of values or worldviews that were developed in the historical crucible of humanity's evolutionary emergence from prehistoric hominids to modern humans. In the Spiral Dynamics model, there are eight in all, although the final two are more speculative, having not entered into the cultural mainstream yet and active primarily in rare individuals. In the 1990s, Beck and Cowan made the unusual decision to color-code these systems, making them more memorable. Spiral Dynamics can be seen as having many parallels to Gebser's model, although it adds more stages. Part of the revelation of an evolutionary worldview is in beginning to see theorists from vastly different contexts tracking such similar territory. The names of the stages may be different, the exact sequencing may vary, but we can see deep and important similarities in the patterns being recognized and the evolutionary dynamics being observed.

Graves's first system (beige), which Beck now refers to as Survivalist, is a sort of clan-based, instinctual, impulsive system in which the goal is to stay alive—an almost pre-language level of consciousness that pre-dates the emergence of contemporary humans. This is similar to Gebser's archaic stage.

The second system (purple), refers essentially to a similar stage as Gebser's magic structure, although it is critical to note Gebser's descriptions are of the system as it first emerged in our ancient past. The basic dynamics of this worldview can still be seen in many indigenous populations, and it is strongly represented in ancient cultures around the world today. Here you have a deeper sense of human bonding as individuals become identified with small clans and tribes; there is a new sense of the dynamics of cause and effect— "the first sense of the metaphysical," as Beck puts it—as early humans try to explain the unpredictable dynamics of their world. Beck expresses that there is a kind of deep human connectivity in this system, a positive heritage of that almost pre-egoic sense of bonding. I saw Beck once play the song "Stand by Me" at a lecture to capture the relational quality of this value system. Here we have animism, and tendencies to think "ritualistically and superstitiously and stereotypically, thus [trying] to control by incantation, totems, and taboos." In this level there is, as Graves describes, a "name for each bend of the river, but none for the river." In the last decades, we have seen a newfound interest in the positive contribution and wisdom of this value system, evidenced by the widespread fascination with shamanism and indigenous cultures.

For the third system of values (red), we have the emergence of the "raw egocentric self—the renegade, the heretic, the barbarian, the go-it-alone, the power-self, the hedonist," as Beck explains. We have the individual self breaking free of the family, the clan, the safe structures of home and hearth. Here we have tribalism in its many forms and ethnocentrism, along with the first empires. In this system of values, we find plenty of rage and rebellion but also creativity and heroism. Think about the microcultures of inner-city gangs and organized crime, but also athletic achievers and rock stars. We might say that there is tremendous positive vitality, energy, and self-expressiveness in this value system.

The next three systems line up roughly with the three most common worldviews active in the world today: *traditionalism (blue)*, *modernism (orange)*, and *postmodernism (green)*. Beck likes to call them Holy Forces, Free Marketeers, and Egalitarians. Think Billy Graham, Bill Gates, and Oprah. Or the religious right, libertarians, and environmentalists. Or Opus Dei, IBM, and Greenpeace. Graves is just one of the tens if not hundreds of theorists and researchers at this point who have identified a roughly similar series of worldviews active today, though it should be noted that Graves was one of the first to clearly identify "postmodernism" or what he called the "relativistic, existential" system. Given that this value system really didn't move deeply into the mainstream until the 1960s, it is understandable that previous theorists might not have identified it so explicitly.

The seventh level, which later came to be called "integral," marked a significant shift, according to Graves, and is a worldview that is as yet unknown in the world, at least on a large scale. At the integral level, values are influenced less by self-interest and more by a desire for the well-being of the whole, the survival and success of the whole project of human existence. Beck describes this system as being "flex flow," a way of acting perhaps best described by Graves:

> The proper way to behave is the way that comes from working within existential reality. If it is realistic to be happy, then it is good to be happy. If the situation calls for authoritarianism, then it is proper to be authoritarian and if the situation calls for democracy, it is proper to be democratic. Behavior is right and proper if it is based on today's best possible evidence; no shame should be felt by him who behaves within such limits and fails. This ethic prescribes that what was right yesterday may not be seen as right tomorrow.

The worldviews of Spiral Dynamics can be thought of as complex systems of values. As Beck described it in an interview, "Spiral Dynamics is based on the assumption that we have . . . complex, adaptive, contextual intelligences, which develop in response to our life circumstances and challenges." They have alternatively been called "value-memes," sometimes abbreviated to "memes" (not to be confused with Richard Dawkins's use of the term*). Ken Wilber, who has incorporated Spiral Dynamics, along with many other developmental models, into his own work, thinks of these stages or codes as "waves of development"—almost like frequency waves across an electromagnetic band, distinct stages that nevertheless blur and blend into one another like colors on the visual spectrum.

---

* Richard Dawkins defined the term as a "unit of cultural transmission" in *The Selfish Gene* (New York, N.Y.: Oxford University Press, 1989), 192.

They are "not rigid levels but flowing waves," Wilber writes, "with much overlap and interweaving."

I think of them as complex intelligent systems that help organize our internal lives in much the same way as the skeletal system or nervous system helps to organize our biological lives. The sheer underlying power and influence of these value systems should not be underestimated, and yet it is important when we think about them to adopt a very flexible attitude, recognizing that they are not rigid or absolute distinctions but significant generalizations that enable us to make much greater sense out of the human experience. They represent deep positions in consciousness, natural attractors that tend to call to themselves ecosystems of values that resonate with the underlying principles of the worldview. Graves believed that they also represented different levels of activation in our neurological equipment, suggesting that these worldviews are not merely psychological and social but neurological.

Beck likes to point out that these value systems represent not so much types of people but types *in* people. We each may express values associated with many of these systems, and yet most people's values will tend to conglomerate around one primary worldview, their "center of gravity," as adherents like to say. It is also wrong to think of these systems as inherently bad or good. They are sets of values that adapt to fit certain life conditions. And there can be healthy or unhealthy behavioral expressions of each system. Extreme cultural relativism would be an unhealthy expression of postmodernism (green) while ecological sensitivity and gender equality would be a healthy version of that same value system. You can have an egomaniac expressing a very sophisticated cultural worldview and a decent, good-hearted person expressing the essential values of a worldview much "lower" on the evolutionary scale. Give me the latter person as a dinner companion any day! We have to be wary of the mind-set that "higher is always better." It's much more complex than that. As Graves eloquently expressed it, "What I am saying is that when one form of being is more congruent with the realities of existence, then it is the better form of living for those realities. . . . When one form of existence ceases to be functional for the realities of existence then some other form, either higher or lower in the hierarchy, is the better style of living." Gandhi's non-violence was a beautiful and effective response to colonialism, but I suspect it would have been less successful in the face of Genghis Khan. Military rule may be an appropriate governance structure for a country on the brink of tribal anarchy, but it would be regressive and disastrous in a more modernized culture.

Those of us who wish to study these stages or transitions through which our consciousness and culture have moved but who lack extraordinary powers of insight into the past need not despair. We can see these stages of cultural development not only by looking back but simply by looking around. As Robert Godwin writes, "We do not need a time machine . . . because in our present world, from the standpoint of psychology, developmental time *is* cultural space." What does this mean? Simply put, it means that individuals and cultures existing today in different areas of the world are at different stages of development. While much of the developed world may have reached modern or postmodern stages, there are many other nations and continents that continue to live at a traditional or even tribal stage. And even within countries like the United States, a number of these different value systems are clearly active at once. Indeed, cultures are never monolithic, particularly in a globalizing world, and within any given country there will be individuals who are inhabiting different stages of development, and moving between them as well.

So while our political system loves to use such distinctions as right versus left or conservative versus liberal as all-embracing categories when it comes to public values, "traditional," "modern," and "postmodern" are actually much better terms with which to analyze social and political movements in this country. For example, when

Richard Dawkins and the new atheists attack religious believers, it's not just atheism versus religion or left versus right. It's modernism versus traditionalism. When scientists attack creationists, it's modernism versus traditionalism. When environmentalists attack the "evil" corporations, it's postmodernism versus modernism. When my parents sent me to a Catholic grade school because the school was a good one, but fretted that their liberal Protestant son might come under the influence of Catholic beliefs, they were worried that the modern values they held dear would be undermined by more traditional ones. My sister and her husband are facing the same dilemma in Houston. Their teenage children attend a private high school run by a Christian organization. The children are intellectually challenged and stimulated by the discipline, high expectations, and value-rich climate of this achievement-oriented school. The effect on their personalities seems positive as well. I'm always struck by how sweet and respectful they are for teenagers, seemingly unaffected by the cynicism, irony-laden attitude, and laissez-faire relativism that they would almost certainly pick up at a more postmodern, conventional high school. And yet when we spend Christmas together and they tell me that their science teacher raises questions about the veracity of evolutionary biology, their uncle gets concerned about the trade-offs involved.

Culture wars are an ancient phenomenon. We can see them in mythology as the "gods" of one value-system battle the "gods" of another. When traditional culture and the monotheistic religions emerged, they struggled for centuries against paganism and polytheism. Many of our current battles have been going on since the emergence of modernism, all the way back in Voltaire's time. Each emergent worldview is, as Hegel told us, in a dialectical relationship to the one before it and is an answer to the problems created by the previous stage. Each one also transcends and includes the values of the previous level of development. The scientific values and achievement-oriented ethos of modernism were a reaction and, in some sense, an answer to problems created by the self-negation, superstition, and otherworldliness of traditionalism, much in the same way religious group solidarity ("anyone can be a Christian") and moral strictures of traditional religion helped mitigate the tribal chaos and ethnic violence of an earlier time period (a developmental process we see occurring again in places like Rwanda). Again, each stage is very dependent on the life conditions of the moment. We postmoderns often lament and ridicule the heavy-handed morality and restrictive attitudes of traditional religion, but if we were raised amid the tribal wars of the Congo, where raping and pillaging have become part of the fabric of a broken society, we might come to appreciate and even embrace the role that restrictive religious attitudes play in establishing strong moral boundaries around the power of unrestrained violence and sexuality. Again, different worldviews are responses to different life conditions. This is not a one-size-fits-all world.

"I WAS ANSWERING QUESTIONS THAT NOBODY WAS ASKING"

Beck was taken with Graves's theory in part because of his own background. In particular, it helped give new perspective to his ongoing contemplation and struggle with that defining feature of American life—race. Beck grew up in Purcell, Oklahoma, and while Oklahoma was never part of the Confederacy or "old South," racial prejudice and segregation was still a major part of life in America's oil patch. Beck recalls a formative moment in 1953 when he was a junior in high school playing a basketball game in the school gym, which had high windows that ringed the court on one side. No blacks were allowed in the gym, but one year, during the intensity of the playoffs, when the locals took a passionate interest in the fortunes of the young team, he noticed faces appearing in the high windows. Wondering exactly what was happening, he walked outside to find out that young black boys were climbing up on high ladders just to get a glimpse of the game. The image of those black boys from the segregated high school across town straining to watch the white team play left an imprint on Beck's mind and brought home the painful and poignant tragedy of America's race relations.

Some years later, this experience would inform Beck at the University of Oklahoma when he chose the Civil War as a subject for his doctoral thesis. Influenced by one of his professors, renowned social psychologist Muzafer Sherif, he studied the process of polarization that made the Civil War an inevitable result of the extreme positions held in this country in the 1800s. He often imagined what he might have done, with current knowledge, to short-circuit the tension of the era. What if he could have had Lincoln's ear in 1859? Could war have been avoided? Little did he know that more than a quarter century later he would get a chance to make a difference in a similar situation.

In 1980, Beck was presenting on Spiral Dynamics at a conference in Dallas. In the audience that day was a South African who had worked closely with both blacks and whites in the coal mines of his native land. "You just explained my country," he exclaimed to Beck after the speech, and promptly invited him to speak at an industrial conference in Sun City. That was the first trip. Beck would make more than sixty in the decade to come, as he fell in love with the unique character of a country where it seemed as if almost all of Graves's value systems were alive and active in the culture, and jockeying for attention and power. "It was a microcosm of the planet," he explains.

When Beck says that all of Graves's value systems or worldviews were active in South African culture, he really means it—with the possible exception of the first (beige) stage, a survival-based worldview usually only seen in desperate circumstances. For example, the magic (purple) worldview was alive and active in the Zulus and their rich culture and sacred places. "The purple worldview is heavily laden with so-called right-brain tendencies," he explains, "such as heightened intuition, emotional attachments to places and things, and a mystical sense of cause and effect. I have a well-developed purple sense myself, having spent so much time with the Zulus." In fact, it was in Africa that Beck began to understand, he told me, the "majesty and dignity" of this value system, one rarely seen in the United States, at least not outside the city of New Orleans or the surviving Native American cultures.

With Graves's work fully internalized, Beck had another critical piece of the puzzle when it came to sorting out the racial prejudices that he had struggled with in his life. Indeed, as he read the newspapers and watched television in South Africa, slowly absorbing the cultural climate and political polarization that was occurring, he realized a surprising truth—one that might have seemed nonsensical to the uninitiated but represented a radically different perspective on the political tensions of the country. "Oh my God," he realized. "This is not about race."

To most South Africans, the societal fault lines were clear. It was black versus white, African versus European. But for Beck, it wasn't so simple. This struggle really masked a deeper conflict, one between value systems. Yes, on the surface it certainly seemed that white Afrikaners simply resented and devalued black Africans and their culture. But according to this new perspective, there was another layer of conflict occurring between worldviews that, in and of itself, had nothing to do with race. So Beck began to educate his audience on the importance of these cross-cultural value systems, pointing out that there were other ways to see the differences among the peoples than through the lens of color. Each race had individuals spread throughout the spiral of development. Not all Afrikaners were the same. Not all blacks were the same. If he could get people to see this, he realized, it would create new pathways for alliances across color lines. Not only did Spiral Dynamics transcend racial distinctions, it had more explanatory power. "And paradigms change only when the new paradigm offers more explanatory power than the one it replaces," he notes.

Beck took up the challenge of South Africa's cultural evolution with a passion of a true believer, the stubbornness of a Texan, and the hardiness of a boy raised less than a decade after the Depression and Dust Bowl. "I had to shape myself to South Africa," he explains. "For example, I respected the Afrikaners rather than condemned them. The only way to speak to the Afrikaner is through religion or rugby, and I chose rugby . . . My role was to shift the categories people were using to describe the South African groupings from 'race,' 'ethnicity,' 'gender,' and 'class' into the natural value-system patterns, allowing for a new dynamic of change. Many were able to connect across these great racial divides to find the basis for a sense of being 'South African.'" He appeared on TV (especially on *Good Morning Africa*, the equivalent to *Good Morning America*), he wrote articles for newspapers, and he inserted himself into every high-level discussion about the future of the country that he could. He made a great many friends and more than a few enemies, some among progressive foreigners who felt he was too accommodating to the white power structure. Beck wanted to find solutions that took into account each of the many worldviews active in the politics of the country, and this didn't sit well with many liberals. "I was advocating a different solution than what the postmodern system demanded, which was the instant redistribution of power, since the only reason for the European-African gaps in development, [according to that value system] was blatant racism."

Exactly how influential Beck was with respect to the transformations that occurred in South Africa and the avoidance of civil war is uncertain. Without a doubt, he was a significant voice among the many contending for power and influence in that country in the late 1980s. What we do know is that the emerging idea of stages and worldviews in the evolution of human consciousness and culture had at last had its day in the political sun. It had gone from the sweeping theories of the Hegelians to the research studies of developmental psychologists to the intuitions of Gebser to the constructions of Habermas, and in the hands of an unexpected advocate it was taken out of the garage of theory and research and allowed to drive around in public. And most important, it played a small but perhaps not insignificant role in avoiding a civil war that would have set southern Africa back generations.

EXPLAINING A COMPLEX WORLD

Frankly, when I first read about Spiral Dynamics, I was unsure what to think. The whole idea seemed so unlikely. "Isn't this oversimplistic?" I thought to myself. "An act of unbelievable hubris and reductionism? How could the extraordinary complexity of human culture be whittled down to essentially eight stages of development? Isn't this exactly the kind of idea that allows us to mistreat and marginalize people of other races, creeds, and cultures?" But as I began to understand the tremendous subtlety and complexity of the theory, I was able to see the truth of it in my everyday experience. And as I identified these cultural codes in myself and in the people around me, and came to appreciate the rich intellectual pedigree of this evolutionary perspective, my initial fears were assuaged. Eventually, I actually began to naturally perceive the world through this spectrum of worldviews. Far from being reductionist, this fundamental idea was enriching my understanding of the human condition. I began to see these differing value sets not as merely good ideas or helpful pointers but as important truths—not absolute truths, not final truths, not scientific truths, but "orienting generalizations," as Ken Wilber likes to say, that help to make profound sense out of the human experience.

Eventually, my fears were turned on their head. I began to see how clunky, ill-advised, and even dangerous it is to act in the world—socially, spiritually, politically, and especially militarily—

with an ignorance of these basic worldviews that structure and condition our lives. Frankly, it's like using nineteenth-century medicine in a twenty-first-century world. Instead of being a vehicle for marginalizing other people and cultures, this perspective is one that I came to see as an essential tool to *prevent* the mistreatment of other peoples and cultures.

Nevertheless, the idea remains controversial, and it will take some time for the understandable stigma associated with stages and hierarchy to work its way through the intellectual currents of our time. Here, again, it helps to keep a certain context. In evolutionary terms, it was not so long ago that human beings were using leeches on the sick or sacrificing babies to appease the gods. One day, I suspect, we will look back and feel similarly about how we have understood cultural development in our own age. That is not to imply that it's easy to make a transition from theory to practice when it comes to the new perspective on evolution. The issues are so complex. Indeed, when it comes to applying the values of one worldview to a society steeped in the values of another, it is easier to do more harm than good. One person's barbarian is another's indigenous elder. One person's genital mutilation is another's sacred ritual. How and where do we make distinctions, draw the lines? And yet the idea that we should unilaterally adopt a hands-off policy when it comes to other cultures is a pretension we can ill afford in today's world.

So we invade Iraq and think that they should be able to immediately embrace the freedoms of modernism and waltz into a democratic future. We're surprised when they don't welcome us with wide-open arms and shocked at the sectarian conflict that erupts. We struggle to understand the nature of tribal dynamics in Afghanistan. Scarred by our failures, we retreat to a live-and-let-live policy, or to a protected isolationism, and hope for peace. Or perhaps we resort to cynicism and embrace pessimistic views of history. Only, that doesn't work either and another Pearl Harbor or 9/11 rouses us from our reverie, calling us to be proactive in the world. But then we imagine in our arrogance that we can quickly remake cultures in our own image, establishing modernism and democracy like some global Johnny Appleseed, distributing our idea of freedom behind the barrel of a well-intentioned gun. And so we alternate between a history-free, naïve idealism that believes too much is possible and a history-laden realism that has no faith in the future. An evolutionary worldview allows us to steer between those extremes and adopt the best attributes of both. Evolutionaries express an idealism that says the future is open-ended and extraordinary change and development is absolutely possible. But they also need to embrace a realism that acknowledges that evolution takes time and that it happens within the context of deep-rooted and complex historical patterns. To simply bypass these, avoid them, or pretend they don't exist is to work in denial of real forces that are shaping the tides of history.

Spiral Dynamics also allows us to get out of the business, as Beck points out again and again, of expecting people to be different from how they are, to somehow change worldviews overnight. That's not the path to pragmatic, workable global cultural evolution—at least not in the short term. "I'm not trying to change people," Beck often says, by which he means he's not trying to manipulate an individual's basic worldview. "People have a right to be who they are." But there are healthy and unhealthy expressions of each code in his spiral system, and some play better with others in our increasingly crowded global melting pot. Indeed, there is a huge difference between the traditional worldview of a Billy Graham and the one of an Osama bin Laden; the modernist spirit that sends people to the moon or the one that turns a blind eye to environmental destruction.

Of course, this cultural perspective is never going to be as simple or as easy to define as individual psychology. Evolutionary systems like Beck's or those of other theorists tracking similar territory certainly do not constitute final proclamations on the nature of human culture. But no longer are they merely pet theories or shot-in-the-dark guesses as to how consciousness and culture evolve. Social science surveys indicate clearly that at least three dominant world-views or value systems are active in the United States. They may not call them traditional, modern, and postmodern, but the data largely corroborate the developmentalists' descriptions. How will we negotiate the dynamics between these worldviews in the years to come? Political pundits often evoke the memories of the good old times of the 1950s and '60s when politicians were more bipartisan and we were able to get more positive legislation through a more amicable Congress. I have my doubts as to the accuracy of their rose-colored memories, but nonetheless, they are right about one thing. The cultural landscape is different today. The postmodern worldview became a force in the United States in the late 1960s and changed the cultural and political character of our country perma-nently. We should focus on understanding the new dynamics of a more complex world rather than longing for a modernist consensus that is lost forever.

Spiral Dynamics, along with other new theories of cultural evo-lution, represent some of the new fruits of the effort to understand this more complex world. But it will be up to future generations to use this emerging knowledge to reshape and transform the contours of our global culture—hopefully, much for the better.

CHAPTER ELEVEN

An Integral Vision

Our first decisive step out of . . . our normal mentality is an ascent into a higher Mind . . . capable of the formation of a multitude of aspects of knowledge, ways of action, forms and significances of becoming. . . . [Its] most characteristic movement is a mass ideation, a system of totality of truth-seeing at a single view; the relations of idea with idea, of truth with truth are not established by logic but pre-exist and emerge already self-seen in the integral whole.

—*Sri Aurobindo*, The Life Divine

Passengers Carter Phipps and Ellen Daly, please report to Gate 14. This is the last call for boarding American Airlines Flight 567 to Chicago."

My wife and I hurtled through the Denver airport, rushing through a special line at security, hurriedly throwing laptops and assorted gear into plastic trays. Finally we were clear, and as we took off across the airport lounge, lunging past restaurants and newsstands, for a moment I felt like O. J. Simpson in those old Hertz commercials, leaping over shoeshine stands in a mad dash to make his flight. I had never been paged in an airport before, and I didn't want to miss the flight—especially now that the entire airport was witness to our drama.

It was a hot summer day in the Mile-High City, and even amid our mad scramble for the plane, I was a happy man. Our tardiness was justified. Our lateness was worth the price. We had just spent several hours in the downtown loft of Ken Wilber, and let's face it—it's not every day that one gets to hang out with one of the great philosophers of our age. American Airlines could wait.

If you're not familiar with the work of Wilber, don't be concerned. He doesn't have a high profile in contemporary culture, choosing to stay out of the limelight and the rough-and-tumble talk-show culture that is the usual path to fame for our public intellectuals. So you won't see him talking to world leaders on CNN. You won't see him schmoozing with politicians at Davos. You won't see interviews with him on *60 Minutes*, *Frontline*, or C-SPAN. You can be an educated, thoughtful, well-informed citizen of the Western world and never have heard his name. But make no mistake, Wilber is important. His work and ideas—he calls it "integral philosophy" but it could just as easily fall under the category of evolutionary philosophy—are quietly affecting the way hundreds of thousands of people think about the world they live in. Ever since publishing his first book, *The Spectrum of Consciousness*, in 1977, Wilber's work has chipped away at the philosophical girders of our postmodern age, clearing out contradictions and confusion and articulating new models and maps of reality that may shape the contours of our future. If postmodernism can be defined, as Jean-François Lyotard famously put it, as "incredulity toward meta-narratives," then Wilber's work is a legitimate antidote. With books called *A Brief History of Everything* and *A Theory of Everything*, Wilber has sought to create the sort of Holy Grail of grand narratives, a vast philosophical framework that attempts to integrate all categories of human knowledge and the history of human cultural development. He attempts to bring together religion, art, morality, economics, psychology, and all of the major sciences under the umbrella of one theory, one meta-perspective. When Wilber uses the world "integral," he really means it.

This was my second trip to his Denver apartment. Wilber, then in his mid-fifties, seemed remarkably fit and full of energy, though I knew the reality was more complicated. Years earlier, he had been exposed to toxic chemicals in a nasty spill near Lake Tahoe, resulting in an autoimmune disease that would occasionally flare up and knock him off his feet for a few days or weeks. Indeed, a couple of years after our visit, the disease would temporarily get the better of him, and a grand mal seizure sent him to the emergency room. He has made an impressive recovery, but while his mind is fully intact, his body presents a daily challenge.

With his tall, chiseled, bodybuilder figure, Wilber sure didn't come across as a bookish intellectual, though there was no mistaking his cognitive gifts. Born in Nebraska and raised a self-described "Army brat" whose family moved around the country, he still has something of that Midwestern hospitableness about him, a warmth of soul that belies his critics' complaints that he is an aloof character, coolly surveying the culture from his mile-high home. On the contrary, he was down-to-earth and engaging, and made both my wife and me feel like close colleagues despite the fact that this was only our second meeting. He does have a marked tendency to curse like a sailor, as if he wants to erase all notions that he's an ivory-tower character.

Wilber's work has taken up the fallen standard of philosophy, the one that reads "The Truth Shall Set You Free" and has wrestled with that great demon of the information age—the explosive proliferation of knowledge to the point of mental overload. Little by little, he has brought forth new, clarifying, and liberating perspectives to help us navigate this chaotic context.

Of course, in a contemporary milieu that has often celebrated feeling and intuition over critical discourse and conceptual illumination, Wilber has certainly earned himself some pointed resentment as well as praise. Moreover, his embrace of evolutionarily inspired notions of purpose and progress, and his placement of spirituality at the heart of his philosophical framework, has not endeared him to a skeptical Western intelligentsia. But even at the beginning of his career, he knew that his own philosophy would run counter to the dominant intellectual currents of the day:

> One thing was very clear to me as I struggled with how best to proceed in an intellectual climate dedicated to deconstructing anything that crossed its path: I would have to back up and start at the beginning, and try to create a vocabulary for a more constructive philosophy. Beyond pluralistic relativism is universal integralism; I therefore sought to outline a philosophy of universal integralism.
>
> Put differently, I sought a world philosophy. I sought an *integral* philosophy, one that would believably weave together the many pluralistic contexts of science, morals, aesthetics, Eastern as well as Western philosophy, and the world's great wisdom traditions. Not on the level of details—that is finitely impossible; but on the level of orienting generalizations: a way to suggest that the world really is one, undivided, whole, and related to itself in every way: a holistic philosophy for a holistic Kosmos: a world philosophy, an integral philosophy.

Wilber is not the first to use the term "integral" in this way. That distinction might better be placed at the feet of Indian sage Sri Aurobindo, or Harvard sociologist Pitirim Sorokin, or perhaps Jean Gebser. Steve McIntosh suggests that they all began using it around the same time in the early twentieth century, unaware of one another's work. Whatever the case, Gebser and Aurobindo have been quite influential on Wilber. Gebser, with his emphasis on the structures of consciousness and culture, helped elucidate the succession of evolutionary unfolding that we have explored in the previous chapters and that Wilber has made central to his philosophy. In other respects, Wilber's body of work resembles Aurobindo's philosophy of "Integral Yoga," at least in its comprehensive evolutionary vision. Of course, Wilber is a philosopher, not a spiritual visionary, and his work incorporates many twentieth-century breakthroughs of which Aurobindo had little knowledge. In particular, Wilber made generous use of the tremendous progress in psychology over the last century, becoming one of the first philosophical voices to incorporate developmental psychology as well as depth psychology into an integral framework.

It would be an exercise in futility to attempt to convey Wilber's full contribution in the narrow confines of a single chapter. Moreover, his work is always evolving rapidly, picking up new streams of knowledge and integrating them into his overarching model, making any definitive statement soon outdated. But I do hope to show how his philosophy helps integrate many of the insights discussed in previous chapters, as well as organize their complexities and contradictions into a more easily comprehensible picture. In these brief pages, I will seek to convey why I believe there is no more powerful theoretical context than integral philosophy to help us understand what evolution actually means in the internal universe.

In this book, I'm calling this new worldview "evolutionary." Wilber, as I've mentioned, prefers the term "integral." Each name emphasizes a different dimension or aspect of this emerging stage of culture, but clearly there is enough overlap here to suggest that these terms are pointing at least in the same direction, if not more or less to the same idea. Is Wilber's evolutionary system *integral*? Or is his integral system *evolutionary*? No doubt both are true. Integral practitioner and author Terry Patten describes integral philosophy as "meta-systemic." It connects the dots, he explains, "and when you connect the dots, the single essential story that emerges out of the otherwise bewildering complexity of our world is *evolution—*

the amazing multidimensional story of our universe's gradually accelerating development—at first cosmic, then biological, then cultural and noetic." Integral, evolutionary, or both, Wilber's novel perspectives on the development of "self, culture, and nature" are seminal for anyone seeking to find their way forward in a postmodern world.

ONE REALITY, FOUR PERSPECTIVES

In the early 1990s, Wilber sat down to write his most comprehensive philosophical work yet, *Sex, Ecology, Spirituality*. It was his first attempt to fashion what could legitimately be called "a theory of everything." It took him three years, and during this extended writing retreat, he ran up against a major problem. The problem was serious but simple—everybody had a different sequence of stages in their developmental system. We've looked at just a couple of such sequences in the preceding chapters, but Wilber was looking at hundreds, in multiple fields of knowledge.

Every school of thought, every system of knowledge, every field of study seemed to have a different set of assumptions as to what constituted the hierarchies of the natural world and of human culture. "There were linguistic hierarchies, contextual hierarchies, spiritual hierarchies," Wilber writes in the introduction to his *Collected Works* (a ten-volume set of impressively weighty books, published in 2000). "There were stages of development in phonetics, stellar systems, cultural worldviews, autopoietic systems, technological modes, economic structures, phylogenetic unfoldings, superconscious realizations. . . . And they simply refused to agree with each other."

As the days and months went by, Wilber sought to bring order to these multiple systems of knowledge. Working alone, in the days before Google allowed instant access to information, he read and read and read. And as he read, he began to make lists of all the various hierarchies that made up the particular structure of any given system of thinking. Those lists eventually begin to collect like stray pieces of clothing all over the floor of his house, a messy collection of data begging for order and clarity, an outward manifestation of a very contemporary conundrum—too much knowledge, too little context. Wilber describes the scene:

> At one point, I had over two hundred hierarchies written out on legal pads lying all over the floor, trying to figure out how to fit them together. There were the "natural science" hierarchies, which were the easy ones, since everybody agreed with them: atoms to molecules to cells to organisms, for example. They were easy to understand because they were so graphic: organisms actually contain cells, which actually contain molecules, which actually contain atoms. . . .
>
> The other fairly easy series of hierarchies were those discovered by the developmental psychologists. They all told variations on the cognitive hierarchy that goes from sensation to perception to impulse to image to symbol to concept to rule. The names varied, and the schemes were slightly different, but the hierarchical story was the same—each succeeding stage incorporated its predecessors and then added some new capacity. This seemed very similar to the natural science hierarchies, except they still did not match up in any obvious way. Moreover, you can actually see organisms and cells in the empirical world, but you can't see interior states of consciousness in the same way. It is not at all obvious how these hierarchies would—or even could—be related.

Archimedes had his moment in the bathtub when he discovered the theory of water displacement; Newton had his legendary moment with the apple; Einstein had what he called the "happiest thought of my life," an insight that led to relativity theory. The genesis of Wilber's own breakthrough insight didn't involve such sudden illumination. In fact, it was more grit than grace (to paraphrase the title of another of his many books). But while Wilber's own epiphany may have lacked the "flash of insight" of those other great moments in the history of human knowledge, don't be surprised if, when the dust clears and we can look back with the dispassionate lens of history, his signature insight compares favorably. Little by little, as he put the pieces together, the answer finally fell into place.

The answer was called "the four quadrants" and it became the basis for Wilber's integral theory and evolutionary philosophy. He realized that all the many hierarchies and systems of knowledge could actually be divided up into four major categories. Some of the hierarchies were referring to collectives, some to individuals. Some were referring to exterior realities, some to interior realities. In fact, it seemed that every hierarchy on his list was referring either to an individual or a collective, and either to interior (subjective) or exterior (objective) dimensions of reality. It all fell together as the various schools and systems separated out into their respective quadrant, and soon he had a clear map—one that contextualized and embraced all of these seemingly conflicting schools of knowledge, one that transcended and yet included the cacophony of completing theories and settled them down into an integrated unity. He had started with a complex, confusing amalgamation of fragmented and isolated data from hundreds of disparate knowledge systems, and he had ended up with one relatively simple, coherent map of the cosmos (or Kosmos, the term he prefers).*

---

* Wilber has reclaimed the Greek term *Kosmos* because, as he explains, "the original meaning of Kosmos was the patterned nature or process of all domains of existence, from matter to math to theos, and not merely the physical universe." (See *Sex, Ecology, Spirituality*, in the *Collected Works* volume 6, page 45.) In other words, he wants the definition of Kosmos to include the internal as well as the external universe.

Ken Wilber's Four Quadrants

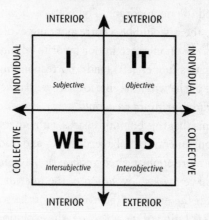

For Wilber, every event in the manifest world can be viewed from any one of those four perspectives: individual interior (I), collective interior (we), individual exterior (it), and collective exterior (its).

Let's take me as an example. I have an *individual, interior perspective*—my thoughts, psychology, and spiritual experiences would fall under that category in Wilber's upper-left quadrant. I also have an *interior collective dimension*, the culture I share with others, including worldviews, values, and belief systems. This is the "we" dimension, the *intersubjective* domain, which corresponds to the lower-left quadrant. Then I have an *exterior, physical dimension*—my brain, body, and nervous system, as well as the objective ways I behave in the world—this is the upper-right quadrant. And I participate in *collective exterior systems*—economic systems, political systems, social arrangements, and so forth, which fall into the lower-right quadrant. And here's the key. Evolution, according to Wilber, happens in *all four quadrants*.

Let's take an example from the biological world: the explosion in hominid brain size that happened over the last several million years. This evolution of the neocortex, according to this map, would not be confined to one quadrant. The physical brain doesn't evolve on its own. So there would be a corresponding evolution in the left quadrants as well, an increase in our favorite hominid's consciousness and capacities (upper-left), and a corresponding change in the culture that he or she was now able to share with others (lower-left). In the lower-right quadrant we would see an evolution in the sociopolitical systems that this hominid shares with others—from primitive tribal arrangements to global economics and trade, for example. In truth, it's a more sophisticated version of the point that so many evolutionists have made—evolution of the interior corresponds to the evolution of the exterior; the evolution of the subjective corresponds to the evolution of the objective; the evolution of complexity corresponds to the evolution of consciousness. Only now we don't have two perspectives from which to examine this process; we have four.

Let's use a more mundane example. Imagine that I go through a religious awakening. I'm born again. I go from being a no-good, petty street criminal to a God-fearing, upstanding Christian. I find Jesus and my life is forever transformed. That's a notable evolution in the self, and it will be reflected in each of these four quadrants. How?

First, that religious awakening is a massive shift in consciousness, and my interior world will be transformed. That evolution will mean that my worldview will shift and it will tend to put an upward pressure on the relationships I share with others. I'll be less likely to hang out with drug dealers and more likely to form a "we" space with those who share my new worldview. The cultural reference points and quality of my personal relationships will be affected, hopefully for the better. Then, in the right-hand quadrants, we would see changes as well. The religious awakening will have effects not just on my consciousness but on my physical brain. New pathways will be created as my behavior shifts and the structure of my gray matter shifts along with it. Researchers talk about "God centers" in the brain. Perhaps I will connect with those aspects of brain chemistry as well. Whatever the case, things will change between my ears, both physically and mentally.

Finally, there is the lower-right, the "its" dimension of evolution. How will that be affected by my religious conversion? Well, imagine the society that is created in the ghettos of the inner city as a result of a poverty-ridden, disadvantaged criminal underclass. That creates its own social system, its own economic system, its own power structures and political order. And they are generally not healthy. But the religious awakening changes that—I begin to participate in and support the creation of new and different social systems, economic ventures, and power structures. My transformation starts to exert an upward, positive pressure on the lower-right structures of my chosen neighborhood. As the internal culture I share with others evolves, so do the external social systems I participate in. As always, my evolution is not mine alone; I'm affecting those around me. We are all, *always*, participating in all four quadrants of reality.

We could have long discussions about which quadrant "leads" the evolutionary process. For Marx, economic structures (lower-right) were the key to cultural evolution. Everything else was "superstructure," meaning that it was sort of a secondary result of changes in economic conditions. For some, consciousness is primary. It all comes down to individuals' shifts in personal consciousness (upper-left). To change the world, they would say, we must first change ourselves and everything will follow from that. But whatever our particular preference, what is perhaps more important is to see how these four quadrants represent a sort of web of connection, a matrix of interconnected structures. Push upward on any one of the quadrants, and the whole dynamic matrix begins to shift. Change in any one area puts positive evolutionary pressure on the other quadrants as well.

Unfortunately, many schools of thought that emphasize one particular quadrant don't even recognize the existence of other schools of thought, and so we get what Wilber likes to call "quadrant absolutism"—people who think their particular corner of reality is the way, the truth, and the light. We see this with certain schools of scientific reductionism—they claim that just about everything can be reduced to the upper-right quadrant. They collapse all the rich realities of the upper-left interiors—thoughts, emotions, psychological structures, spiritual experiences, and so on—into the upper-right ("It's all just brain chemistry"). Or with schools of Marxism, everything is reduced to economic structures, the lower-right quadrant. And we also see it with schools of mysticism or idealism, which claim that everything can be reduced to consciousness, the upper-left quadrant. We see it with philosophies that hold that all behavior can be reduced to deep social and cultural influences, the lower-left quadrant. This basic insight has been a guiding light for Wilber's work—that most schools of thought are neither right nor wrong but rather "true but partial," meaning they are right in their own limited corner of the universe but fail to take into account the larger ocean of ideas in which they swim. Thus, even in being right, they can be horribly blind. Even in being true, they can be partial. "The real intent of my writing is not to say, You must think in this way." Wilber explains, "My work is an attempt to make room in the Kosmos for all of the dimensions, levels, domains, waves, memes, modes, individuals, cultures, and so on ad infinitum. I have one major rule: *Everybody* is right. More specifically, everybody—including me—has some important pieces of truth. . . . But every approach, I honestly believe, is essentially true but partial, true but partial, true but partial. And on my own tombstone, I dearly hope that someday they will write: He was true but partial."

The four quadrants were instrumental in helping Wilber make sense out of the mess on his floor and the knowledge in his mind. And while they have remarkable breadth, they also have another important dimension: depth. Wilber has been one of the most prominent advocates of the core idea we have been exploring in this section, that individual and collective evolution progress through a series of discernible stages of consciousness and culture. And he is perhaps the first to notice just how much common ground there is among theorists and researchers from vastly different fields—from the social evolution stages of Habermas to the psychological stages of Baldwin to the cognitive stages of Piaget to the moral stages of Kohlberg and Gilligan to the cultural stages of Gebser to the sociological stages of Paul Ray, the ego-development stages of Loevinger and Susanne Cook-Greuter, the media theory stages of Marshall McLuhan, the spiritual stages of Sri Aurobindo, the color-coded cultural stages of Spiral Dynamics, and many, many more.

Gathering together all of these various systems under one roof, it became clear to Wilber that while each of these developmental, evolutionary theories is tracking different specifics, there is a clear evolutionary pattern at work. We are not looking at hundreds of different systems of development, all disparate and disconnected. Despite their varying disciplines, contexts, and fields, they have a remarkable amount in common. In fact, Wilber noticed, they display a consistent series of seven or eight clear and distinct worldviews, waves, levels, or stages, viewed from a number of different perspectives.

He began to call his map "AQAL"—all quadrants, all levels. Only by embracing all of these truths, only by acknowledging all of these important facets of reality, Wilber came to believe, could we even start to have a reasonable, non-distorted discussion about human knowledge, human culture, and how to address the many problems of our world. For Wilber, this is the test of any perspective in an integrally informed, evolutionary age: Does it include an awareness of all quadrants, all levels of reality—at least all of those we have collectively explored thus far in the human journey? And while Wilber has never shied away from the hierarchy that such a schema implies, he also recognizes, along with Gebser, Graves, and most sophisticated developmental theorists, that as we progress upward along the path of evolution, we must be careful not to draw facile moral conclusions about what it all means. He writes:

> [T]his does not mean that development is nothing but sweetness and light, a series of wonderful promotions on a linear ladder of progress. For each stage of development brings not only new capacities but the possibility of new disasters; not just novel potentials but novel pathologies; new strengths, new diseases. In evolution at large, new emergent systems always face new problems: dogs get cancer, atoms don't. Annoyingly, there is a price to be paid for each increase in consciousness, and this "dialectic of progress" (good news, bad news) needs always to be remembered. Still, the point for now is that each unfolding wave of consciousness brings at least the possibility for a greater expanse of care, compassion, justice, and mercy, on the way to an integral embrace.

"In every work of genius," wrote Emerson, "we recognize our own rejected thoughts; they come back to us with a certain alienated majesty." Wilber's work is no exception. His theory has that unique trait of explaining the world in a way that seems completely novel yet somehow familiar at the same time. Once you deeply internalize the power of this simple map of reality, it lodges in your mind like a mental filtering mechanism, a natural organizing principle that begins to clarify and contextualize incoming data almost before thought. Any attempt to reduce the evolutionary process to one dimension of that four-part map begins to feel like a dangerous form of fundamentalism—not religious, perhaps, but limited and problematic nevertheless—a distortion of reality that inevitably leads to all sorts of conundrums and confusions, the disastrous result of which we see splashed across the headlines every day of the year.

A NEW PERSPECTIVE ON EVOLUTION

"The course of evolution has, just on this planet, gone from dirt to the sonnets of William Shakespeare. How can anyone not see meaning in that?" Wilber exclaimed to Ellen and me in his apartment that day in Denver, his voice rising as he delivered a mini-discourse on the current state of evolutionary theory. "How does one miss these extraordinary higher stages of unfolding? How could one look at dirt on the one hand and the music of Bach on the other and say, 'Oh, they're both equally meaningless phenomena.' What kind of person could come up with that conclusion?"

Wilber is indeed passionate about the meaningful nature of evolution and has devoted a significant part of his philosophical work to understanding exactly how evolution works in the deeper dimensions of self, culture, and nature. As I have mentioned before in this book, most Evolutionaries see the nature of evolution through a particular lens, reflecting their particular angle on the deeper dimensions of reality. Some talk about the evolution of genes, some the evolution of cooperation, some the evolution of consciousness, some the evolution of information, some the evolution of values, some the evolution of economics, some the evolution of empathy, and so on. While these approaches can be complementary, they nevertheless each reflect the fundamental viewpoint of their proponents. In this sense, Wilber's evolutionary vision is no different. While he is attempting to integrate many of these various currents of thought into one framework, he also privileges a very particular approach to the evolutionary process as being perhaps more true, more accurate, more foundational. So what is the unit of evolution that he sees as being the most essential element of this cosmic process? *Perspectives.*

Wilber tracks the evolutionary journey as a development of perspectives. His integral philosophy "replaces *perceptions* with *perspectives*," as he puts it. It "re-defines the manifest realm as the realm of perspectives, not things nor events nor structures nor processes nor *vasanas* nor *dharmas*, nor archetypes, because all of these are perspectives before they are anything else, and they cannot be adopted or even stated without first assuming a perspective."

It is a powerful and novel way to think about evolution. And it points to capacities we take for granted today that may not have always existed as potentials in consciousness. For Wilber, each evolutionary step in culture, each new wave of consciousness, each new worldview—from tribal to traditional to modern to postmodern, or from egocentric to ethnocentric to worldcentric—is also a fundamentally new perspective, a new vantage point from which to see the world.

To illustrate this, I like to think of the iconic picture of the Earth from space. That picture, in all its many forms, has helped to give us an amazing new perspective on the life of our planetary system. Beyond its aesthetic beauty, that picture represents something important, a sense of unity perhaps, an awareness of the system of which we all are a part, of our common home and the fate we share on this remarkable blue-green planet. It almost represents a new capacity in consciousness, the ability to embody the perspective, in some rudimentary way, of the planet itself. We can look at things from the perspective of how they influence, benefit, affect, or harm the overall integrated planetary system and all of its inhabitants. We can step beyond our personal perspective or the perspective of any particular species and grasp, at least temporarily, a critical viewpoint that has nothing to do with our everyday concerns. Not everyone alive today is able or willing to take such a perspective. In fact, it is still a rare capacity, and during other times of human history it simply was not possible to think about life in that manner. Such a planetary perspective was probably not even available to human consciousness, except perhaps through some remarkable individual cognitive leap. That is not to denigrate our ancestors. Rather it is to point out that we perpetually underappreciate our evolutionary achievements. And a big one of those achievements is the capacity to embody more encompassing, more comprehensive, and more integrated perspectives.

There is another, perhaps more humbling, implication of this "ontology of perspectives." It also means that we are already always embedded within a perspective. In the actual interior world of my own consciousness, I am not just a subject apprehending other objects in consciousness. I, the subject, am also embedded in very real perspectives that often remain unseen by the interior vision of my mind's eye. In other words, I may think that I have some objective relationship to the events in my perception, that I am seeing reality "as it is," but that is a trick, an illusion of first-person consciousness. My awareness is always conditioned by a perspective. It is an insight that runs counter to a fundamental assumption of the meditative and contemplative traditions—that knowledge gained through introspection is trustworthy. Wilber points out that postmodern thinkers such as French philosopher Michel Foucault accurately noticed that this was a fallacy endemic to introspective traditions. We are always embedded in intersubjective cultural structures that influence our "objective" perception of reality. For example, if a Christian monk has a vision of Jesus, he may believe that he is seeing an objective spiritual reality. But he fails to recognize, these thinkers tell us, that his vision—even if it is spiritually authentic—is inevitably being influenced by tremendous cultural and social conditioning that is all taking place prior to and outside of the monk's immediate awareness. Like a Hindu witnessing a vision of Krishna, or Tibetan experiencing a powerful visitation from a bodhisattva, he is mistaking a cultural archetype for objective perception.

In the preceding chapters, we explored how unseen intersubjective worldviews and structures of consciousness condition our awareness more than most people realize. Many of us are unwitting mouthpieces of subtle but incredibly influential worldviews—whether we like it or not, whether we think we are or not. We may think that we perceive reality as it is, but in fact we are each more like the protagonist in our own personal version of *The Truman Show*, and we cannot see the subtler cultural forces that are invisibly shaping all of our perceptions—even our most cherished spiritual experiences.

So what's the good news? The good news is that even our capacity to understand this truth shows evolution at work. We are gaining the capacity to take more and more broad, encompassing, sophisticated, and subtle perspectives. Look how far we've come in the last fifty thousand years. Once upon a time, we could likely do little but appreciate and apprehend our own vital experience, but slowly we developed unprecedented new perspectives. Take, for example, the capacity to "step outside of ourselves" and think about our own experience with some greater measure of objectivity. That's an extraordinary leap forward in evolutionary terms. More recently, we've developed methods for exploring the natural world with the tools of science. That too is a perspective—one that has disembedded itself from the natural world enough to begin to analyze it. This capacity was not always part of the human experience but rather was the hard-won result of the evolution of perspectives over time.

Eventually we developed the capacity to stand in another person's shoes, to imagine, through an extraordinary leap of consciousness, what it might be like to feel their emotions, to understand their suffering, to appreciate their joys and sorrows, even in circumstances vastly different from our own. This is yet another great gift of evolution, and one that not every individual on the planet has actually developed. And evolution continues. We are developing ever more subtle perspectives—for example, the capacity to take the perspective of the species itself, to imagine what might be good or bad for the future of humanity as a whole. We have even developed the capacity to take the perspectives of other species, to begin to imagine the quality of their consciousness and adapt our human activities to start to take into account their existence. And now, in just the last few decades, we are beginning to develop, as we have seen in this book, the extraordinary capacity to take the perspective of evolution itself, to step inside the mind of the process, so to speak, and explore reality from that very remarkable and revelatory perspective. This has just happened—in an evolutionary blink of an eye. So don't underestimate the power of perspectives and the effect of their evolution on every last dimension of our lives.

THE KOSMIC GROOVES OF EVOLUTION

Thinking about Ken Wilber put me in mind of another underappreciated American philosopher. Charles Sanders Peirce was celebrated by his contemporary William James, praised as the greatest philosopher in America by Bertrand Russell, declared the American Aristotle by Alfred North Whitehead, and respected by countless other intellectuals of the nineteenth and twentieth centuries. Yet his legacy is relatively obscure, and he is barely recognized for his contributions to evolutionary thought. Like Wilber today, he was, in his own time, one of the most important thinkers the average educated American citizen had never heard of. And when it came to evolution, his philosophy was equally innovative. In fact, Peirce set in motion a way of thinking about the subject that would, more than a century later, inspire one of the aspects of Wilber's work that I find most useful to understanding an evolutionary worldview.

Peirce lived a strange and in some respects tragic life. Born the son of a Harvard professor in 1839, he was a successful student, but history tells us that his interpersonal skills—his emotional intelligence, we might say—didn't quite match his IQ. He seemed to suffer from a difficult personality (some of which stemmed from a physical condition) that earned him many powerful enemies and complicated his efforts to obtain academic respectability and credibility. But whatever his physical and psychological limitations, his mind lived on another plane, and his voluminous writings (mostly unpublished in his lifetime), his eclectic brilliance, and his ground-breaking thought were many decades ahead of their time. Though Whitehead only became aware of Peirce's work more than a decade after the philosopher's death, he recognized in this evolutionarily inspired pioneer a kindred spirit and saw important similarities to his own work. Like Whitehead, Peirce was at the forefront of a change in the way we understand the universe, seeing evolution, movement, and process where others saw only fixed laws, dead matter, and eternal stasis.

I found it remarkable to discover, in the course of my research, that all the way back in the nineteenth century Peirce was questioning the spell of solidity even as it applied to the most sacred cows of the physical sciences: the laws of nature. For Peirce, the entire universe and all of its forces and creations were subject to evolution. Indeed, Peirce's work was one of the first to begin to theorize how something as ostensibly absolute as a law might be created through the processes of evolution. Perhaps the laws of nature are not unchanging, applying to everything for all time, he suggested. Perhaps they didn't pre-date the universe. Perhaps they, too, evolved along with the forms and structures of our cosmos.

Peirce suspected that many of the seemingly fixed structures of our universe are in fact better described as *habits*—habits that have become so deeply embedded in nature that they behave like laws, fixed and unchangeable. In 1915, the *Mid-West Quarterly*, a publication of the University of Nebraska, published the following description of Peirce's ideas as presented in his lectures at Johns Hopkins University.

> May not the laws of the universe be the acquired habits of the universe? May there not still be a possibility of the modification of these habits? May there not be the possibility, forever, of the formation of new habits, new laws? May not law be evolved from a primordial chaos, a universe of chance? In the play of chance

still apparent may we not see the continual renewal of the life of the universe, a continual renewal of the capacity for habit forming and growth?

Peirce suggested all of this before science had any sense of cosmological evolution, of the deep-time developmental history of our universe. Questions concerning the laws of physics are even more fascinating today, particularly in the context of our current understanding of Big Bang Theory. Did those laws exist in some timeless void prior to the initial cosmic emergence? Did they pop into existence at the moment of that great conflagration? Were they gifts, perhaps, of a previous universe, a sort of cosmically inherited informational DNA designed to help structure the evolution of our own realm of time and space? When it comes to such issues that get at the heart of our cosmic origins, we still have far more questions than answers.

Biologist Rupert Sheldrake is another thinker who has suggested that the laws of nature may not be immutable and eternal but are more like habits. And he points out that most physicists have not thought deeply about these questions in light of our new cosmology. "Although cosmology is now evolutionary," he writes, "old habits of thought die hard. Most scientists take eternal laws of Nature for granted—not because they have thought about them in the context of the Big Bang, but because they haven't." Lately, it seems, a few more physicists have stepped into the breach with interesting speculations about the source of the laws of nature, such as Templeton Prize winner Paul Davies and science writer James Gardner. But wherever such speculations ultimately lead us, what is important for our discussion is that once again the spell of solidity is broken and we can at least begin to consider the possibility that certain characteristics of the universe that seem immutable and unchanging might better be considered as evolutionary—things that develop over time through habitual repetition until they become more and more established. Eventually, in a cognitive illusion that fools us again and again, they seem fixed, eternal and unchanging, when actually they are nothing of the sort. Again, my point is not to weigh in on questions of physics or make any claims about what happened in the far-off reaches of deep time but rather to illuminate a way to think about evolution that becomes particularly interesting when we apply it—as Wilber would do—to the internal universe.

But first, Sheldrake would attempt to apply it to biology, with a theory he called morphic resonance. In 1981, he published *A New Science of Life*, suggesting that a new kind of information field, a "Morphogenetic" or "morphic" field, may be critical in helping to determine the forms and structures of living systems. A morphic field is a place where information is stored, where the informational memory or "habits" of past forms and structures reside and influence the form and content of the present. "According to this hypothesis," Sheldrake writes, "systems are organized the way they are because similar systems were organized that way in the past. For example, the molecules of a complex organic chemical crystallize in a characteristic pattern because the same substance crystallized that way before; a plant takes up the form characteristic of its species because past members of its species took up that form; and an animal acts instinctively in a particular manner because similar animals behaved like that previously."

We can see a very nonesoteric version of this "repetition creates habit" idea in the remarkable plasticity of the brain. Our contemporary appreciation of neuroscience tells us that when we act in a novel way, we are creating new connections in our brains that support that particular new behavior. Repeat the behavior and the neural connections are strengthened. Each time we act or think in that particular way, the neural pathways become a little more established. The habit is slowly forming. We are developing new pathways in the brain, new habits of mind, which in turn correspond to new ways of thinking and living. Obviously, there are constraints on how much the brain can change, but the basic point stands: novel behaviors create new pathways that develop into habits, and perhaps eventually into instincts.

In the 1980s Sheldrake's proposals were summarily dismissed by the scientific powers that be and his book was referred to as the "best candidate for burning there has been in many years" by one of the more respectable scientific journals. This is in part because there is no place in existing biological theory for nonphysical, or very subtle, information fields that influence the development of biological form (though, as Sheldrake points out, the field of physics is much more open to such possibilities). He has recently republished the original book with updated data and has proposed a number of testable experiments to support his ideas and carried them out with mixed results. He has won some converts but has done little to convince skeptics.

Whatever the scientific fate of the idea in physics or biology, once again, the evolutionary principle is important—as Wilber has recognized. He has suggested that Sheldrake's morphic fields are similar to deep structures of consciousness in the *internal* universe. Thus, the stages of development that we have discussed in previous chapters become, from this perspective, psychological, social, and cultural "grooves," as Wilber likes to call them— "Kosmic habits" that have developed in the noosphere. Citing the common observation that once a certain task, such as synthesizing complex molecules, or rats learning a particular maze, has been accomplished in one part of the world, it can more easily be accomplished somewhere else, Wilber draws a parallel to the emergence of psychological forms: "In historical unfolding, once . . . [a stage of development] had significantly emerged anywhere in the world, it began more easily appearing elsewhere around the world. A difficult, novel, creative emergence had settled into a Kosmic habit."

To illustrate Wilber's point, I like to imagine I am walking into an unexplored land. In order to get from point A to point B, I must choose a particular route. In my wake, I leave a path, but a barely distinguishable one, only a slight imprint of my footsteps that a subsequent traveler might notice. Indeed, the second traveler may or may not see that trail I followed, but we might say that there is a slightly increased chance that he or she will be drawn to travel along the same pathway. With each subsequent traveler, that likelihood increases; as the path gets more well worn, the memory gets more stable, the groove in the land cuts deeper. Eventually, after millions of individuals have passed that way, the trail, now a well-established pathway, acts as a powerful attractor to any traveler on the same route. It would seem almost nonsensical not to follow it. One may add slight features to the existing path, but when it comes to the basic trail, the structure is relatively stable and mostly unchanging, a strong mold set into the internal universe. In fact, it might seem as if it were a permanent "unyielding mold" that has always been there. Such is the nature of deep Kosmic grooves. We may not be able to see them the way we can see a trail in the ground, but they are carved deep into our consciousness nonetheless, and they exercise the same power of attraction over us. And the older the pathway, the older the structure, the more settled and determined it is.

Perhaps the best part of this perspective is that it explains the endurance of the past and its influence on the present, but like all good developmental theories these days, it is also upwardly open, allowing for the development of new habits and evolutionary grooves as humans evolve forward in history and confront new challenges.

> Most of the early stages of [human] development have been around for thousands of years. And billions of human beings have gone through them so that now they are automatically part of development. They're as rutted as the Grand Canyon. . . . But new stages . . . might be a yard or two deep, that's all that's been cut yet. And

so, boy, it's hard to make things stick in that. And anybody who's pushing into those stages is basically going out next to the Grand Canyon, taking a stick, and starting to dig another groove. . . ."

Wilber further explains:

This does not mean that individuals cannot pioneer into these higher potentials . . . only that those structures are as yet lightly formed, consisting only of the faint footprints and gossamer trails of highly evolved souls who have pushed ahead, leaving gentle whispers of the extraordinary sights that lie before us if we have the courage to grow. These are higher *potentials* and nonordinary *states*, but . . . they have not yet become *structures* settled into stable Kosmic habits.

Let's take, for example, the evolutionary worldview that this book is endeavoring to elucidate. At this point it remains unformed, a fascinating mixture of new ideas and like-minded little memes and values, all connected to larger breakthroughs in our understanding of evolution. To pick up on my earlier image, it's multiple sets of footprints, all heading in roughly the same direction, crossing over one another, sometimes rambling, not yet a clear trail but a compelling imprint. Wilber refers to this time as the "frothy, chaotic, wildly creative leading-edge of consciousness unfolding and evolution, still rough and ready in its newly settling contours, still far from settled habit." And he makes a critical point: "This is why today, right now, we want to try to lay down as 'healthy' a . . . groove as we possibly can, because we are creating morphic fields in all subsequent Kosmic memory."

This brings to light another important implication of the notion of Kosmic karma or evolutionary grooves—it points to a new kind of moral imperative. As we begin to recognize ourselves as participants in the creative process, we also realize that there are ethical implications to this perspective. What we do *matters*—not just for its effect on the present but for its effect on the as-yet-unformed future. We are not just participating in an already-created universe; we are also participants in the process of creation. And that knowledge carries tremendous moral weight. It comes with a very real ethical context, Wilber suggests—one based not on some decree from an all-knowing God or religious belief system but on our emerging understanding of the way in which evolution works and the role we play in that process. Are you ready to be the individual who walks out on that virgin land and takes responsibility for laying down a positive trail? Are you willing to be the person whose behavior, good or bad, positive or negative, evolutionary or deevolutionary, could ratchet into the fabric of the universe a new habit, a new groove, one that may very well influence others? In *Enlighten-Next* magazine, Wilber pointed out:

> Paul Tillich said that what we call the Renaissance was participated in by about a thousand people. And that is astonishing. About a thousand people defined an entire culture by the choices they made at the leading edge and because they were choosing from the highest stage of development at that point, they were laying down structures that became the future of humanity in the Enlightenment, and then in modernity and then in post modernity. A thousand people. And the same thing can happen today if you are awakened to the leading edge . . . you very well could be part of the next thousand people that are laying down the form of tomorrow. . . . And we're doing this now as a conscious process.

Now, before I start planting seeds of grandiosity, let's understand that it is not every day that Joe Normal is going to be walking around defining new grooves in the internal universe. Most of us live out our entire lives in the well-worn grooves of yesteryear.

Nevertheless, the basic point stands. Wilber calls this emerging moral sense the "evolutionary imperative," reminiscent of Immanuel Kant's categorical imperative. Kant was famous for his suggestion that each person should behave as if his actions would become a universal rule. We can see in his categorical imperative the dawning of a powerful world-centric morality, a sense of the universality of our ethical concerns. Human beings in his day were just beginning to think about morality in a new way, breaking out of the smaller ethical context of tribes, nations, or religious faiths, and starting to consider the moral truths that concern humanity as a whole. And Kant was leading the charge. Two and a half centuries later, the idea is reemerging, only this time it is not a world-centric morality we are talking about, it is a Kosmos-centric, evolutionary morality that is now the leading edge of our own discussions of how to redefine right and wrong in a postmodern culture that is none too crazy about either of those notions. Like almost everything else in an evolutionary universe, morality evolves. And with this new imperative, Evolutionaries find themselves embedded once again in a powerfully ethical context, connected to the newly discovered truths of an evolutionary universe, awake to the deep interior consequences of our actions, and consciously responsible for the choices we make at the edges of the future—where novel habits form, and new trails, lightly traveled, may slowly become the accepted pathways for tomorrow's culture.

PART IV

REENVISIONING SPIRIT

CHAPTER TWELVE

Evolutionary Spirituality: A New Orientation

*Once men thought Spirit divine, and Matter diabolic. . . . Now
science and philosophy recognize the parallelism, the approxi-
mation, the unity of the two: how each reflects the other as face
answers to face in a glass, nay, how the laws of both are one. . . .
  We are learning not to fear truth.*

—*Ralph Waldo Emerson, "The Sovereignty of Ethics"*

In February of 1836, the world's most famous scientific voyage
docked in King George Sound in Southwest Australia on its way
to circumnavigating the globe. The ship, named the *Beagle*, pulled
into the harbor, and its young passenger, Charles Darwin, disem-
barked and spent eight days in this lonely outpost of the British
Empire. One day, while he was visiting a local settlement, a large
tribe of native Australian aboriginals stopped by to trade with the
colonialists. And after exchanging some goods with the English set-
tlers, Darwin reports that these natives held a great dancing party
in the evening, a sort of tribal ritual. His accounts make it sound

something like a rave at Burning Man, with less Ecstasy and more spears. Our young evolutionist was not impressed, calling it a "most rude, barbarous scene."

"Everyone appeared in high spirits," he wrote, "and the group of nearly naked figures, viewed by the light of the blazing fires, all moving in hideous harmony, formed a perfect display of a festival amongst the lowest barbarians."

Almost two centuries later, Nicholas Wade, the science editor for the *New York Times*, wrote in *The Faith Instinct*, his book on evolution and religion, that these "emotionally compelling dramas of music, chant and dusk-to-dawn dances" represented some of the very first forms of religious expression. Darwin's dancing tribesmen were a modern window into an ancient rite, a glimpse of how our distant ancestors paid tribute to a sense of the divine.

Personally, I find this evolutionary tidbit fascinating and more than a little ironic, because one of the earliest memories I have of the influence of religion on my own life is watching my sister struggle to understand why her date to the biggest dance of the year, where she was to be crowned queen of the school, would not be allowed to dance with her because of his parents' religious beliefs. Her anger and tears etched in my ten-year-old mind the painful consequences of religious conservatism. I guess the hardline Baptists in my hometown hadn't heard the news about Darwin's natives and the roots of religious faith.

In the vast historical epochs between the let's-dance-all-night tribal mind of ancient humans and the Jesus-hates-dancing conclusions of some contemporary Christians, we find quite the range of religious expression. It's so great in fact, that to even begin to treat it all as one phenomenon would seem an exercise in futility. Nevertheless, that hasn't stopped a growing wave of Darwinian scholars like Wade from trying their hand at explaining the evolutionary logic of why some people bend their knee to exalt a supreme being and others boogie all night by the campfire to celebrate a higher power.

When I marry the words "evolution" and "spirituality," this is what comes to mind for many—the effort to explain the evolutionary origins of religion and the spiritual impulse. There are in fact, as we shall see, other profound and interesting ways to combine these two terms, but perhaps the most common way is to start asking questions like: Why did religious faith evolve? What accounts for the emergence of the "faith instinct" or the "God gene," to use the current terminology? According to some scholars, religion must have evolved because it provided some kind of adaptive advantage to our ancestors. As Kenneth V. Kardong, author of the recent *Beyond God* bluntly puts it: "Religion is not inserted by the hand of God, nor is it an outgrowth of the human psyche attempting to deal with the realities of existence. . . . Like large brains and upright posture, religion arose for the survival benefits it bestowed." Perhaps in the extraordinary group cohesion that the religious impulse fosters among believers, the thinking goes, we can find the key to its adaptive advantage. Religious solidarity, which generally is considered to be one of the strongest social glues that we know of, even today, could be the hidden motive behind the persistence of faith in human history. After all, few agreements form the kind of tight bond that religious agreements do. And those bonds may have been crucial to enduring the trials and tribulations of life over the course of the history of our species. Religious faith endures today, we are told, not because it represents any kind of accurate or well-conceived appraisal of how the universe works but because it was those groups with religious faith that survived and thrived and therefore passed on the particular instinct to their descendants—us.

It's a good hypothesis, and I suspect there may be seeds of real truth in it. Unfortunately, such accounts tend to be used not simply to help us understand the multidimensional reality of the human spiritual and religious impulse but to oversimplify it and explain it away, to dismiss it entirely as a sort of industrious and happy delusion that tricks us into fulfilling a positive social purpose. Granted, that's probably a significant upgrade from Marx's view of religion as the "opiate of the masses," but we should be extremely careful about such sweeping conclusions. They tend to be more ideological than empirical. Luckily, we can always embrace the fascinating insights of new knowledge and research without adopting the attendant reductionism. In other words, we can appreciate the legitimate and fascinating science that helps fill out the picture of how something like religion evolved without necessarily accepting the oft-added overlay of philosophical and metaphysical conclusions that better reflect the attitude of the author than the logical conclusion of the research.

There is another problem with such thinking. It tends to paint religious and spiritual pursuits with one generalized brush, not fully accounting for the incredible diversity of what we call religious practice and the way it has changed and developed through history. We might say that this is partially the result of an impoverished understanding of the development of consciousness and culture. In such accounts, there is very little sense of the evolutionary view of the emergence of human consciousness that we've been exploring in the previous chapters of this book. And without that perspective, the Flintstones fallacy runs rampant. The moment we accept the possibility that the human mind has changed quite radically over the course of history, then we need to rethink our relationship to religion. And then we can begin to see the obvious truth that religious and spiritual pursuits have also changed dramatically over the course of the last ten thousand years—arguably in tandem with the development of human consciousness.

Indeed, it is critical that we break the spell of solidity in the way we think about religion. As worldviews change and evolve, so too does religion. This is another way in which the terms "spirituality" and "evolution" are being combined. There are not only various cultural expressions of religion in different parts of the globe; there are *types* of religious expression that correspond to certain levels of individual and cultural development. As Wilber, Beck, and others have pointed out, the way we interpret any spiritual or religious experience is going to change over time, depending on the worldviews that condition all of our experience. A person whose developmental center of gravity is at the mythic or traditional level is going to have a very different type of Christianity from a modernist, and that in turn will be quite distinct from a postmodern orientation. Think about the difference between the warlike, tribal Christianity of the Crusades; the mythic rituals of traditional Catholicism; the achievement-oriented modern message of megachurch pastors like Joel Osteen; and the peace-loving, interfaith-oriented approach of Unitarian churches. Not only are these obviously very different interpretations of the faith, they are also developmental stages in its expression that correspond to the cultural stages described by Gebser, Graves, and others.

We can find a similar notion explored in Emory University professor James W. Fowler's work on the "stages of faith." A colleague of Lawrence Kohlberg at Harvard, Fowler developed a theory of stages of religious development, which he outlined in his 1981 book, *Stages of Faith*.

Fowler's work examines religion, in the broadest sense of that complex term, as the way in which we think about and relate to the ultimate nature of existence. Under that definition, even many forms of modern atheism and naturalism are really better classified as modernist forms of religious expression (Stage Four in Fowler's construction). After all, they often represent strong conclusions about the ultimate nature of being upon which whole systems of cultural rules, norms, and ethics are constructed. And it's important to note that in Fowler's sequence, as in all the more sophisticated developmental systems, higher is not always better. There can be healthy and unhealthy expression of each stage of evolution's unfolding.

Once we truly begin to appreciate the evolutionary nature of even a universal phenomenon like religion, we can begin to see how regrettable it is that so many scholars, especially scientists, tend to think about it as a single phenomenon, as if most of human history can be broken down into a simple two-step affair. First there was religion, then science. First faith, then reason. First belief and superstition, then logic and rationality. First supernaturalism, then naturalism. In such formulations, all forms of religious expression get lumped into one broad category. That is a misleading way to think about religion, because important distinctions, such as those discussed above, will be overlooked and smudged together, leading to inaccurate conclusions about the whole subject.

Current debates about God often find themselves embroiled in these distortions. The New Atheists, despite their laudable championing of modernity's gifts of science, reason, and rationality, tend to propagate this unfortunate confusion. While some may be quite clear about how they define religion, associating it narrowly and exclusively with a faith in a supernatural, mythic God (or gods)—a faith that is still the foundation of many of the belief systems active in the world today—many are less careful. They tend to see all mystically or spiritually inclined individuals as being afflicted with more benign strains of the same underlying disease. What they often fail to acknowledge is that not all religious expression is created equal. Even under the umbrella of any particular tradition, there are vastly different ways of thinking about God, each of which represents certain worldviews, perspectives, and stages of faith. And for those of us who enthusiastically embrace both the deepest intimations of the spiritual impulse *and* the tremendous virtues that flow from the project of science, the first order of business is to free the idea of spirit from being frozen in history and exclusively associated with the traditional, mythic, transcendent, otherworldly, anthropomorphic, dogmatic, old-man-in-the-sky-God belief system, by whatever name.

The terms "evolution" and "spirituality" meet in this endeavor to understand the evolution of the religious impulse, as well as in the search for the god gene or faith instinct. But there is yet a third confluence of these ideas, a form of specifically *evolutionary* spirituality that is emerging in our time. Rather than simply looking through an evolutionary lens at already established forms of spirituality or religion, some Evolutionaries are intuiting and forging a new spirituality, a new theology, a new mysticism, a new cosmology, and a new morality that is an expression of the evolutionary worldview I've been sharing in these pages. Not simply accepting of the idea of evolution, it is informed by the insights and perspectives revealed by our relatively new knowledge of our cosmic, biological, and cultural origins. This new spirituality is distinct from both traditional religious theism and the anything-goes "spiritual but not religious" pluralism more common in progressive culture.

Evolutionary spirituality is evolution-inspired, world-embracing, and future-oriented. It is a creative, anticipatory spiritual path in which salvation, however we define that word, is to be found not in connection to the ancestral spirits of yesteryear, in promises of a heavenly beyond, in achieving a transcendent state of inner peace, or even in letting go into a timeless present, but in fully embracing the emergent potential contained in the depths of an evolving cosmos.

FOR THOSE WHO LOVE THIS WORLD

He was a child of Italian immigrants, raised Catholic in a poor neighborhood in Queens, New York, in the 1930s. His mother and father never learned to read, but it would be a book that changed his life. A smart kid, he earned a scholarship to a good Catholic prep school, but religious questions began to trouble him. By the time he got to law school, his confusion had grown. Unable to reconcile himself to the basic tenets of the faith, he was searching for answers.

Then in 1958, Angelo Giuseppe Roncalli was named Pope John XXIII, and something important changed in the Catholic faith. A Jesuit author, once banned, had died a few years earlier, and his books—long the subject of rumors, whispers, and shocked speculation—were released from theological censorship and made available to the laity. The young man wasted no time. He sought out the works of this dangerous thinker and began to read. It didn't take long for the effect to be felt. When he opened the covers of a book titled *The Divine Milieu*, the first six words—the book's dedication— hit him like a lightning bolt: "For those who love this world."

The young man was Mario Cuomo, and he would go on to become governor of New York. The source of his inspiration, of course, was that patron saint of Evolutionaries: Teilhard de Chardin. When Cuomo recounted this story to me some years ago, I found it remarkable—not only because of the deep influence of Teilhard's evolutionary thinking on an individual who once seemed close to the American presidency but also because his experience is an archetype for the shift in religious revelation that an evolutionary worldview makes possible.

Cuomo, like millions around the world, was raised in the kind of religious context that holds that the primary goal of the religious life is to be found beyond this world—whether it be in the joys of a heavenly afterlife, the bliss of nirvana, the perfection of the Buddhists' "Pure land," or the transcendent peace that "passeth all understanding." In such traditions, this world—its joys and sorrows, pleasures and vices—is considered to be a dangerous place that will test your soul and tempt you away from the true paradise awaiting the pure and righteous.

"The rule of the Catholic Church was: If you enjoy it, it's a sin. If you enjoy it a lot, it's a mortal sin," Cuomo explained to me. "The world is a series of moral obstacles. And the mission is to avoid temptation, avoid ambition, avoid functioning too much in this world—don't get used to it, because the real joy comes later on in eternity. That was religion."

The more traditional the context, the more starkly we can see this basic message. In some strains of religious thought, modernity has done much to moderate this impulse, but make no mistake: the antiworld bias that was part and parcel of religion for thousands of years is still alive and well. It is easy to see it, of course, in the afterlife promises of the Christian tradition or the Islamic faith, where all kinds of strange enticements—from bliss, happiness, and moral righteousness to sex with virgins—have been used to keep believers' attention less on the day-to-day and more on the hereafter. The Eastern traditions of Buddhism and Hinduism, with their visions of multiple lifetimes and reincarnation, have not altered the basic theme—the goal is *moksha* (spiritual freedom) or enlightenment or nirvana, forms of liberation that free the soul from the shackles of time and the bondage of birth and death. Even the Buddhist ideal of the bodhisattva, so popular in "spiritual but not religious" circles, is merely a different chord in the same basic song. The bodhisattva represents a spiritually liberated individual who vows to delay his or her own liberation until all sentient beings are free from the shackles of time and rebirth. Selfless, for sure, but still in service to an otherworldly ideal.

To be sure, there are many ancient and contemporary spiritual and religious teachings that present themselves differently—non-dual mystical paths that claim no fundamental difference between form and emptiness, between spirit and matter, between heaven and earth. But dig a little beneath the surface and more often than not, the subtle antiworld bias will reveal itself. There are exceptions, such as the world-embracing traditions of the Jewish faith, where the glory of God is revealed in our attempts to perfect and heal the world; in lineages like Kashmir Shaivism, whose tenets specifically distinguished themselves from the world-denying tendencies of Hindu thought; or in the *via positiva* traditions that have emerged from time to time in the development of Western thought. But the deep marriage of spirit and an otherworldly form of transcendence is still unquestioned by most.

Any definition of a new evolutionary spirituality starts with the breaking of this bias. Evolutionary spirituality, though it comes in many colors, has a message much more suited for the life conditions of the modern and postmodern world: The *evolution* of this world is the goal of spiritual life. And by "world" I mean the manifest cosmos of time and space, both the interior and exterior realms—consciousness, culture, and cosmos. The action is *here*—in this time, in this place, in the possibilities that lie in the near and distant future of this culture, this world, this universe. Yes, there still may be spiritual transcendence of the most radical, sublime, and subtle forms, but *transcendence is in the service of evolution*, not the other way around. And that difference is everything.

Transcendent states of peace and freedom, subtle spiritual realizations, liberating insights, and mystical awakenings are not in and of themselves the goal of evolutionary spirituality. They may be authentic, profound, and life changing. But ultimately, the freedom, insight, and liberation they confer is in service to a larger context, a greater project, a higher goal—the evolution of self, culture, and perhaps, if we may be so bold, of the cosmos itself. That is not to diminish the beauty and majesty of transcendence in all its myriad forms but to expand the circumference of its transformative effects far beyond the confines of any individual self. Evolutionary spirituality calls on us to participate in the deeper processes at work in the development of culture and cosmos, and the experience of transcendence, in this context, must ultimately point us forward—not upward, downward, or inward.

Perhaps the two individuals who brought this distinction most powerfully to light were Teilhard de Chardin and Sri Aurobindo. One in the West; one in the East. Both lived in the first part of the twentieth century, during what we might call the first wave of evolutionary spirituality. Both were presenting a new vision of their own faith—Teilhard an evolutionary Catholicism and Aurobindo an evolutionary yoga in the Hindu tradition. In order to make their case, each attacked the antiworld bias of his own tradition. And they did so with ferocity.

Teilhard's thought has surfaced again and again in these pages, a testimony to the breadth of his influence on all facets of an evolutionary worldview. But as we move into the territory of evolutionary spirituality, his own life story becomes relevant as an example of the birth of a new religious sensibility. Teilhard's world-embracing sensibilities were nurtured among the verdant, wooded hills of Auvergne, France. Born in 1881, he spent his formative years just a few miles from the distinctive marker of that region, the massive Puy de Dôme volcano. And though the fire of this long-dormant natural wonder had grown cool over the preceding ten thousand years, the energy of the land began to nurture in Teilhard's young heart and mind a different kind of crimson glow—what he later called the "divine radiating from the depths of blazing matter." It was this unusual passion for the creative energies and potentials contained in matter that characterized Teilhard's temperament from the start. He loved nature, but not with the aesthetic sensibility of a Romantic. He intuited a deeper dimension to the natural world, sensing that contained in matter was a powerful latent potential that was somehow in process—moving, developing, and building to some future culmination.

The profoundly optimistic and future-oriented thought of this Evolutionary was forged in an unlikely crucible: the battlefields of World War I. A volunteer stretcher bearer who repeatedly chose to stay on the front lines rather than accept a safer job as reward for his valor, he used his four years on the brink of death to contemplate the deeper currents of life. Gazing out one night at the battlefields of Verdun, lit up by flares but now quiet, he wrote: "As I looked at this scene of bitter toil, I felt completely overcome by the thought that I had the honour of standing at one of the two or three spots on which, at this very moment, the whole life of the universe surges and ebbs—places of pain but it is there that a great future (this I *believe* more and more) is taking shape." Later, he would reflect on the experience of the front lines: "You seem to feel that you're at the final boundary between what has already been achieved and what is striving to emerge. . . . The mind . . . gets something like an over-all view of the whole forward march of the human mass, and feels not quite so lost in it. It's at such moments, above all, that one lives what I might call 'cosmically.' "

Returning from war, Teilhard was a decorated hero, but for him, the real medals of war were the sheaf of essays, letters, and other writings that contained his newly crystallized ideas. The Church authorities, however, were not amused, and he was forbidden to publish. An early essay/poem titled "Hymn to Matter" makes it obvious why. It expresses exactly how this passionate young Jesuit felt about the world of nature and, by extension, the doctrines of the Church:

> Blessed be you, harsh matter, barren soil, stubborn rock. . . .
> Blessed be you, perilous matter, violent sea, untameable passion.
> . . . Blessed be you, mighty matter, irresistible march of evolu-
> tion, reality ever newborn; you who, by constantly shattering
> our mental categories, force us to go ever further and further in
> our pursuit of the truth. . . .
>
> I bless you, matter, and you I acclaim: not as the pontiffs
> of science or the moralizing preachers depict you, debased,
> disfigured—a mass of brute forces and base appetites—but as
> you reveal yourself to me today, *in your totality and your true
> nature*. . . . I acclaim you as the divine *milieu*, charged with cre-
> ative power, as the ocean stirred by the Spirit, as the clay molded
> and infused with life by the incarnate Word.

Teilhard's tumultuous life took him halfway around the world to China, at the behest of the Church, where he would pen his most famous manuscript, *The Phenomenon of Man*, as well as a host of essays and meditations while living as a kind of exile. Besides his philosophical musings on evolution, he would make his mark on science as well, participating in some of the most significant paleontological discoveries of the day. Frowned on by the Vatican and rarely welcome in his home country, where his thoughts and lectures tended to stir up too much excitement for the Jesuit hierarchy, he was initially known more for his scientific work than his spiritual thought, and this remained true up until the time of his death in 1955 in New York City. In fact, a very ambivalent relationship with evolution—much less with "evolutionary spirituality"—prevailed in corridors of Rome until the mid-twentieth century. Henri Bergson's popular *Creative Evolution* was outlawed by the Vatican and placed on their list of banned books, a heretical distinction that even Darwin's *On the Origin of Species* never managed to achieve.

On the other side of the world from Teilhard's homeland, India's great evolutionary sage Sri Aurobindo (whose equally colorful life we will explore more fully in chapter 16), made a similarly trenchant critique of the antiworld bias present in much of the Buddhist and Hindu traditions. He was certainly not the first to do so. Western commentators had for years made various claims that Buddhist philosophy equates to some form of nihilism or that Eastern mysticism rejects the material world. But they had done so with their own Western religious biases and from an outsider's vantage point, often with little true knowledge or experience of the powerful mystical insights, spiritual realizations, and enlightened states of consciousness that the Eastern traditions offer. Aurobindo was different. While schooled in both East and West, he was a native Bengali, and his own spiritual awakening came in the context of a yogic practice. A defender of the tradition, he knew it both philosophically and experientially.

In his spiritual masterpiece, *The Life Divine*, written in the early twentieth century, Aurobindo presents two "negations" that haunt the contemporary mind. The first is the "materialist denial," essentially the belief that matter is the be-all and end-all of reality. Aurobindo rejects this position and argues for a "more complete and catholic affirmation" of reality in which the material and immaterial realms are both included. Overall, however, he maintains a positive attitude toward the secular turn in global culture and praises the "indispensable utility of the very brief period of rationalistic materialism through which humanity has been passing." He acknowledges that spiritual and religious aspirations often lend themselves to interpretations of reality that strain contemporary credulity, and that the embrace of a skeptical, modern attitude could ultimately serve to facilitate a deeper, more "austere" and authentic relationship with spiritual matters than had previously been possible.

> For that vast field of [spiritual and religious] evidence and experience which now begins to reopen its gates to us, can only be safely entered when the intellect has been severely trained to a clear austerity; seized on by unripe minds, it lends itself to the most perilous distortions and misleading imaginations and actually in the past encrusted a real nucleus of truth with such an accretion of perverting superstitions and irrationalising dogmas that all advance in true knowledge was rendered impossible. It became necessary for a time to make a clean sweep at once of the truth and its disguise in order that the road might be clear for a new departure and a surer advance. The rationalistic tendency of Materialism has done mankind this great service.

The second of the two negations is what Aurobindo calls the "refusal of the ascetic." And the sage saves his sharpest barbs for this topic. By "the refusal of the ascetic" he is referring to the opposite tendency to that of materialism: the refusal to accept that the material world is real. In a groundbreaking and courageous chapter, he attacks the antiworld bias of his own tradition with all the intensity of a man who has seen the mountaintop and knows beyond a shadow of a doubt that it is only a stepping-stone to the next summit.

He begins with a full acknowledgment of the profound experiential basis for this tendency toward an antimanifestation, world-is-an-illusion transcendence that infuses so many traditional religious attitudes. He does not reject the ascetic spirit. On the contrary, he notes its "indispensability" and its ongoing contributions to human advancement. But he explains that there is a type of experience that can easily lead us to the conclusion that the world is an illusion or somehow is not real, and we have to be very careful about how we interpret this kind of mystical realization.

> For at the gates of the Transcendent stands that mere and perfect Spirit described in the Upanishads, luminous, pure, sustaining the world but inactive in it, without sinews of energy, without flaw of duality, without scar of division, unique, identical, free from all appearance of relation and of multiplicity—the pure Self of the Adwaitins, the inactive Brahman, the transcendent Silence. And the mind when it passes suddenly, without intermediate transitions, receives a sense of the unreality of the world and the sole reality of the Silence which is one of the most powerful and convincing experiences of which the human mind is capable.

Under the powerful influence of this type of experience, Aurobindo explains, it is easy to draw inaccurate conclusions about the nature of the spiritual life and the right relationship to the manifest world. He suggests that the dangers here are "parallel" to the materialist denial of spirit but "more complete, more final, more perilous in its effects on the individuals or collectivities that hear its potent call to the wilderness." In other words, we can get lost in matter and materialism, but we can also get lost in spirit and a kind of idealism in which only *that* is real—consciousness, spirit, Brahman, and so on. For Aurobindo, this second negation is the more dangerous and powerful delusion.

Of course, as we sit here and reflect on these issues amid the material comforts of the twenty-first century, almost a century after Aurobindo wrote those words, it may seem strange to imagine that embracing an antiworld spiritual bias could be the more enticing of his two "refusals." After all, just take a look at the recent spiritual book and movie phenomenon *The Secret*, where the salvation on offer was not heavenly bliss but a decidedly material fulfillment. There is little doubt that whatever the dominant tendency of early twentieth-century India, here in the twenty-first-century West, materialism is the dominant "refusal" of the day. Even most religious pursuits in the new millennium don't exactly cry out with a passion for asceticism. And yet we should not dismiss the great sage's warnings as outdated and culturally irrelevant.

The "refusal of the ascetic" is not simply a stance of outward renunciation. Aurobindo is also talking about an inner position, and here the story gets a little more interesting. Even on the progressive edges of culture today, where ancient religious dogmas hold little sway, the tendency toward a kind of transcendence that countenances an ambivalent relationship with the manifest world is alive and well, albeit in somewhat subtler forms. Look, for example, at the most popular Western mystic of our day, Eckhart Tolle. Tolle's book *The Power of Now* inspired millions, including the great oracle of modern media, Oprah Winfrey. A powerfully awakened mystic, Tolle writes in *The Power of Now* that "the ultimate purpose of the world lies not within the world but in transcendence of the world." He once declared in an interview with *EnlightenNext* that the manifest world is simply "ripples on the surface of being" and explained that "every phenomenon in the manifested is so short-lived and so fleeting that, yes, one could almost say that from the perspective of the unmanifested, which is the timeless beingness or presence, all that happens in the manifested realm really seems like a play of shadows." Tolle's image echoes a long religious and mystical tradition. "Most theology takes God to be the only real thing there is, all else being only shadows on Plato's cave," writes religious scholar Huston Smith.

WHY MATTER MATTERS

When I was twenty-two, I spent a month in Katmandu, Nepal. I was there to visit the beautiful Himalayan nation and to participate with some friends in a month-long retreat. I have many fond memories of that trip to the legendary "roof of the world"—beautiful picnics high in the terraced foothills, morning glimpses of the towering peaks before smog and clouds closed in, quiet evening meditations on the balcony, and visits to the local Buddhist temples. But it was also the scene of my first face-to-face encounter with this subtler yet powerful version of an antiworld bias. It came in the form of a Tibetan Buddhist monk.

The monk in question was a student of a revered Tibetan teacher, a man whom many considered to be one of the highest living expressions of the Buddhist teachings at this time in history. He graciously visited us one afternoon, and we ended up having a long discussion on the roof of our hotel. In many respects, we saw eye to eye on spiritual matters. He too was a serious practitioner and we traded stories and perspectives as to the role of practice on the path to enlightenment. It was hard not to be impressed with the sincerity and dignity of this impressive Tibetan. But when it came to the role of thoughts and emotions in living a spiritual life, something changed. Suddenly, we weren't so together after all. Every time I referred to a positive emotion—a sense of love, joy, conviction, enthusiasm, good motivation, spiritually inspired excitement, and so on, he would dismiss it immediately. "All display!" he would say with a wave of his hand. In other words, all of these emotions, no matter what their actual content, were nothing but illusions in our minds, empty expressions of the eternal polarity of fear and desire, paths to inevitable attachment and suffering. The best response was to stand back from all of one's experience like a passive, watchful Buddha, unperturbed by the flux of emotions as they dance across the perceptual screen of the mind.

There is, of course, a spiritual potency to this type of orienta-tion, especially when it is backed up by genuine experience. Indeed, anyone who has ever tasted the deep peace of meditation, come to rest in the stillness beyond the mind's endless chatter, glimpsed freedom from desire and release from the illusion of material needs, knows for themselves the power of this "pathless path." And they can understand at least in some measure why Aurobindo called it "one of the most powerful and convincing experiences of which the human mind is capable." As he suggests, many are so convinced of the reality of the transcendent experience that they conclude that the things of this world are more or less an illusion—the play of samsara, the false world in which unawakened beings suffer and are reborn until they attain enlightenment or nirvana and find eternal release.

Traditionally, such conclusions inspired acts of asceticism, a lit-eral rejection of society and its material trappings. Today, however, there is no great likelihood that the average spiritual seeker will find him or herself compelled to give away all money and possessions, abandon home and family and social responsibilities, and set out barefoot for whatever remaining wilderness can be found. Like the conclusions that lead to it, the dangers today of an antiworld bias are far subtler. But the consequences remain perilous. Especially in an irony-laden, angst-ridden age, we need not be wandering ascetics to become charmed by the seeming spiritual legitimacy of a perspec-tive that allows us to observe rather than act, witness rather than engage, and remain deeply ambivalent about life rather than take the risk to wholeheartedly commit. Such an approach to spirit can easily lead us to the conclusion that what happens in this world is not of great importance. This insidious inference sneaks in around the edges of our minds, undermining not only our sense of conviction in the consequential reality of our actions but also our appreciation of the possibilities that the world presents to us. In a cynical, disbelieving culture, where self-doubt is endemic and so many are hesitant to deeply invest themselves in life and in the possibilities of the future, such conclusions only fan the flames of the wrong kind of detachment. And in the name of peace and spiritual freedom, they can serve to disconnect us from the life-affirming potentials that, as we will see, are the essence of evolutionary spirituality.

Indeed, there is nothing we need more right now in our culture than the kind of conviction that comes from knowing at a deep level that what we do is real and genuinely matters, that it has ethical import. And I don't just mean ethical in the sense of some ultimate cosmic moral tabulation but in the sense that our actions are connected, in some small but significant way, to the fate of a process much larger than ourselves. Evolutionary spirituality can awaken us to that connection. It can link us to a universal process that is not neutral or inconsequential but more real and important than anything that happens in the up-and-down vagaries of our individual lives. It can lift us beyond the endless cycles of our psychological concerns and connect our own capacity for choice to a 13.7 billion–year process, one that is playing out in the evolution of human consciousness and culture in our time. It can reinvigorate our faith, as I pointed out in chapter 3, in the possibilities of our individual and collective future. And it can make us true believers again—not in heavenly deities or godly favors but in the positive, deeply spiritual nature of life and human potential.

Indeed, one of the many functions of spirituality and religion throughout the ages has been to connect human beings to a sense of something beyond ourselves, a greater context, a higher order. In *The Real American Dream*, Andrew Delbanco, director of the American Studies program at Columbia University, argues that this urge to transcend the confines of the self is a fundamental part of human nature, which has, over time, shifted its focus but not lessened in intensity. In fact, he writes, "the most striking feature of our contemporary culture is the unslaked craving for transcendence. . . . I see no reason to doubt—and I do not think history supports such doubt—that human beings of all classes and all cultures have this need for contact with what William James called the 'Ideal Power' through which that 'feeling of being in a wider life than that of this world's little interests' may be reached. . . . The question we face today is how, or whether, this 'feeling of being in a wider life' is still available."

I would suggest that today the "feeling of being in a wider life"— whether we call it God or transcendence or even enlightenment— is most certainly available. In fact, it is much more so, because we no longer need to look to an otherworldly ideals to find awe, wonder, humility, and moral courage that this greater context provides. James's "wider life" has become tangible in the extraordinarily vast spatial context that we have now discovered to be our universe, and the unfathomably deep temporal context within which we have found ourselves situated. The limits of our "little interests" are shattered in awakening to the reality that we are much more than merely an infinitesimal speck in that infinite, impersonal expanding cosmic sea. Rather, our cosmic home is not just a place but a process; *it is moving*, and we are at the very forefront of that temporal unfolding, awakening for the first time to both the physical and spiritual dimensions of that truth—digesting the implications, coming to terms with its existential significance.

Almost a century has passed since the evolutionary spiritual visions of Teilhard and Aurobindo emerged, along with those of their many great contemporaries, such as Bergson and Whitehead.

Yet while science, culture, philosophy, and spirituality have certainly evolved since their day, the underlying perspectives these great pioneers offered still feel surprisingly fresh. Perhaps these Evolutionaries of the past were treading a path that their own time and culture was not ready for. While their footprints were striking and profound enough to have remained visible to this day, few were ready to follow and lay down a path, a "Kosmic groove" into the future. But today, this seems to be changing. The evolutionary visions they intuited seem to be finding a greater resonance at the beginning of our new century than they found in their own time. As more and more Evolutionaries awaken to this world-embracing, future-oriented spiritual sensibility, the contours of the path are becoming discernible. And while broad enough to encompass many variations, the direction of the path is clear, and its spiritual, moral, and philosophical imprint on contemporary culture is deepening. In this section we will explore some of these extraordinary emerging visions of evolutionary spirituality.

CHAPTER THIRTEEN

Conscious Evolution: Our Moment of Choice

The evolution of man is the evolution of consciousness and "consciousness" cannot evolve unconsciously. The evolution of man is the evolution of his will, and "will" cannot evolve involuntarily. The evolution of man is the evolution of his power of doing, and "doing" cannot be the result of things which "happen."

—*Gurdjieff,* Letter to Ouspensky, *1916*

It was a moment of choice. Lying alone in a Sydney hospital room in 1951, Zoltan Torey was on the verge of death, facing a decision that would determine not only *if* he lived but *how* he lived. A twenty-one-year-old Hungarian refugee, he had come to Australia to escape a life under communism, and was working in a local factory at night to finance his days at university. A few evenings earlier, disaster had struck. While hauling a large drum of battery acid across an overhead track, the plug came loose, and a deadly shower of corrosive liquid rained down. "The last thing I saw with complete clarity was a glint in the flood of acid that engulfed my face," he wrote years later in his autobiography *Out of Darkness*. "I recall reeling back, gasping for air with my nose, mouth, and eyes full of the stuff, coughing and spluttering . . . I spun around, beginning to notice a fast-thickening fog passing over my eyes. At this point my consciousness exploded in a sense of catastrophe. There was no thought in that instant, just fragments, faces of people dear to me, and a sickening feeling of this being the end. Then the fog closed in."

When he woke up the next morning in the hospital, Torey was blind and would never see again. His vocal cords were heavily damaged, and he could speak only in a whisper. But that wasn't the worst of it. The true casualty of this industrial accident was something deeper in the self. Yes, the acid had burned through his cornea and ravaged his throat, but it had done equal damage to his soul. All of his dreams, his plans for the future, the possibilities that make a twenty-one-year-old get up in the morning, had been washed away in that one splash of pain and fate. As he lay in the hospital room in the days that followed, his once-bright prospects looked grim. "I virtually witnessed my life sinking, ebbing away," he remembers.

At the height of his struggles, feverish and dangerously depressed, Torey's will to live was clearly failing and everyone knew it. He remembers a concerned hospital staff member sitting down next to him, and suggesting: "Maybe you should pray."

Maybe he should pray? The thought struck Torey, lodged in his mind like a problem he couldn't solve, and as night descended on the Sydney hospital, he contemplated his situation. Yes, he was in a mess. He needed help. He knew that he might never walk out of that hospital. But pray to God? Nominally raised as a liberal Lutheran, Torey was not personally religious. In fact, he had never prayed before. What right did he have now to ask God for special favors? And even if some distant deity could hear his cries, how did it make sense, in a suffering world, that divine favor should be bestowed on him? In 1951, the world seemed to be one disaster area after another, with war, famine, and diseases still threatening much of postwar Europe and Asia. And here he was, in a hospital room, thinking God should be concerned with his personal pain just because some random act of fate had intervened in his life? Why should providence give some special dispensation to the newly blind Zoltan Torey? *Maybe he should pray.* He understood the need but just couldn't bring himself to do it. It seemed the height of selfishness.

As the night deepened and the hospital quieted, Torey reflected on his own situation. What if, instead of asking God for help, he reversed things? Perhaps the more important question was not how God would help him but how he would help God? And not even God, really, but life—this process we are all a part of. How might he contribute—help further the things he deeply cared about, such as life and love and understanding and clarity?

In order to answer that question, Torey began to review what he knew about life, the universe and, well, everything. He began to reflect upon the entire sweep of cosmic history. He had always been interested in truth and science and finding the deeper meaning and mechanics of things, but now in the midst of crisis, that inner curiosity took on an urgent and focused intensity. He reviewed in his own mind what he knew about the evolution of the universe—the singularity that started it all, the breakthrough of life, how chemistry became biochemistry, and how consciousness began. And it struck him—the clear directionality of it all. Matter had become conscious. Configurations of infinitesimal atoms had over eons of time become autonomous thinking and feeling human beings. How extraordinary! The obvious conclusion stared him in the face. *Human consciousness is the advancing edge of this magnificent process.* And this has implications, he realized. There is a moral code embedded in that recognition. We are not just part of this process, but we *owe* it our interest, effort, intellect, and cooperation. We are caught up in this incredible momentum, and we carry the meaning of the process forward into the future. Evolution has given us remarkable brains and a powerful form of consciousness, and in return it is our privilege—no, our *obligation*—to promote and contribute to the betterment of this process. We are not victims. *We have a choice*.

A new path had unfolded in front of his sightless eyes, and Torey recognized that his life still had great potential for meaning. He could still add to the evolutionary process, to enhancing life and furthering human knowledge. "It would justify me and my life and my youth, the privilege of being alive," he recalls thinking.

Torey was saved that night. Maybe not by a supernatural deity answering a pious prayer, but make no mistake—he woke up a changed man, touched by the power of an evolutionary epiphany. Making his blindness an asset, he went on to use his heightened powers of mental visualization to further our understanding of how consciousness works in the human brain, a contribution that has won him respect from some of the greatest minds in science. His 1999 book, *The Crucible of Consciousness*, was one of the more sophisticated attempts to model how the brain creates the experience of consciousness and how the miracle of self-reflection has evolved in the human animal. Torey is also a deep humanist, someone whose infectious love for life and for the miracle of evolution shines only that much more brightly given the backdrop of his personal tragedy.

Torey's story is inspirational but also instructive. It represents a personal victory over the temptations of despair but also an impersonal dawning of a new context in which to see the important choices of human life. It shows how a new kind of ethical context has quietly descended on human awareness over the last years and decades, almost unrecognized, like an unseen snow falling softly during the night, only to reveal a changed world in the light of the morning sun.

Torey is part of a loose category of thinkers I'll refer to as the "conscious evolutionists." In a broad sense, it is simply another term that points to the new worldview that all Evolutionaries in this book embody and represent. More specifically, it refers to those individuals who see in life's current trajectory a critical moment for our species, a time when we must consciously seize the reins of our destiny, take control of the evolution of human society, and face the future with a new sense of agency and choice. And for each of these thinkers, the word "conscious" is critical. They feel that, like Torey in that hospital bed, we must *wake up* and choose to lend our conscious support to the most important endeavor there is—the evolution of our species. And just as Torey reached a critical moment in his personal confrontation with despair, so too, these individuals suggest, may we as a species be reaching an inflection point in the evolutionary trajectory of human culture. Conscious evolution is the need, capacity, and urgency to take responsibility for the future of this experiment called human life. Barbara Marx Hubbard, the influential evolutionary activist who popularized the term in her 1998 book, *Conscious Evolution*, describes it as "a vision and a direction to help us navigate through this transitional period to the next state of human evolution."

There are certain Evolutionaries who embody this calling with unusual passion and inspiration, and whose life stories, like Torey's, serve as examples of this particular kind of *awakening to evolution*. In this chapter I'll take a look at the lives and work of three such individuals—the Evolutionaries who have together probably done the most to bring the transformative idea of conscious evolution to a mass audience—Michael Dowd, Barbara Marx Hubbard, and Brian Swimme. Or, as I like to think of them—the Preacher, the Matriarch, and the Cosmic Bard.

THE PREACHER

"Humanity is the fruit of fourteen billion years of unbroken evolution, now becoming conscious of itself," declares the middle-aged man as he paces back and forth, punctuating his points with a dramatic gesture or a momentary pause. The reverend is in his element, and today he can feel that the crowd is in the palm of his hand.

"When the Bible speaks about God forming us from the dust of the Earth, it's metaphorically true," he exclaims, articulating his words like a verbal challenge. "We did not come *into* this world— we grew out from it, like an apple grows from an apple tree. That statement from Genesis is a traditional way of saying the same thing. We are not separate beings *on* Earth, living *in* a universe. We are a mode of being *of* Earth, an *expression* of the universe."

Dressed in nondescript slacks and a conservative button-down shirt, Michael Dowd reminded me of the Christian ministers of my youth: the wholesome, boyish looks; the clean-cut aura; the warm, inviting smile that whispers of faith and conviction; the natural sense of connection with his audience, be it one person or several hundred. And of course, there's the passion.

"Do you get this?" he asks the audience, eyes bright, searching around the room for response. "We are the universe becoming conscious of itself. We are stardust that has begun to contemplate the stars. We have arisen out of the dynamics of the Earth. In the words of physicist Brian Swimme, four billion years ago, our planet was molten rock, and now it sings opera. Let me tell you, this is *good news*! And I love talking about it!" The last words come out as a shout, and he jumps up to add emphasis, overcome by his own exuberance. The crowd at this mid-sized venue in Cambridge, Massachusetts, laughs, enjoying this unusual preacher of an unusual gospel, although old-time Pentecostal-style passion wasn't what they expected when they signed up for an evening lecture on the "Epic of Evolution." And the evening is just getting going.

There have been tougher crowds for Dowd, venues where he was lucky if he could convince the audience that the dinosaurs *didn't* die out five thousand years ago in Noah's flood. Today's audience— liberal, open-minded Boston intellectuals—is a little more the norm. Dowd is an author, speaker, minister, and self-styled "evolutionary evangelist." He and his wife, science writer Connie Barlow, were traveling together around the country, itinerant missionaries evangelizing evolution on the highways and byways of America, preaching to anyone interested in hearing the "good news" of a life informed and enriched by the "Great Story" of our cosmic evolutionary heritage. It is a story that started in the world of science but, according to them, is destined to transform the world of spirit.

It is said that those who are most passionate about religion, or for that matter about almost anything, are those who convert, not those who are born and raised within it. The most zealous teetotaler is the former alcoholic, the most passionate Christian is the converted sinner, and in this case, the most inspired advocate of an evolutionary worldview is the former antievolution fundamentalist. Believe it or not, there was a time when Dowd would more likely have been the heckler in the audience warning of the satanic evils of evolutionary theory. "I was once one of those people that you see passing out those antievolution tracts," he admitted. "I would argue with anyone who thought the world was more than six thousand years old." He paused, then smiled. "So whatever your name for Ultimate Reality is, he, she, or it obviously has a sense of humor."

Born on the latter end of the boomer generation, Dowd was raised Roman Catholic, but his teenage years were marred by an early drug addiction. Eventually, deciding he needed a fresh direction, he dropped out of college and headed over to that most iconic place for new beginnings and life changes—the military recruitment office. Soon he was on his way to Germany as a newly minted military police officer. As it turned out, drugs were just as easy to come by in Germany, and Dowd's struggles continued. Eventually, he was busted, stripped of his command duties, and demoted to infantry. Lost and confused, he drifted for a while longer until finally one day, backpacking on a mountain overlooking Frankfurt, he came upon that other renowned dealer of fresh starts and new beginnings—God. He also had a little help from an old friend.

"I smoked some good dope and I had the most profound mystical experience of my life," he told me. "I felt that God was asking me, 'What are you doing with your life?' I had a vision that I lived ninety years and I died. The question haunted me: 'What difference did I make? What changed?' The answer I got was that maybe the mountains would have eroded a little. If there was a major flood the rivers might have changed course. Ultimately, what is the significance of a person's life? This is what I felt God was saying to me: 'I want you to make a difference. And it was then that I realized that I was so addicted, I was in deep trouble."

The next Sunday, Dowd made it to church for the first time in many years. The minister was showing a Billy Graham film and at the end of the service, as was customary, he asked if there was anyone who wanted to come on down and give their life to Jesus Christ. Dowd was as ready as they come. He bolted to the front and had a born-again experience. Overnight, he became, as he puts it, "a passionate, radical Jesus freak," and for him personally, it was truly a godsend. He stopped doing drugs and cleaned up his life.

But transformation came at a price. As Dowd's life changed and his heart opened up, his mind closed down. He had jumped headlong into one of the more conservative, fundamentalist strains of Christian life, and as a newly saved convert, he was ready to give it his all. He took on the tenets of his new faith with great fervor: antievolution, the Second Coming, Satan, hell and damnation, the end times. He fasted, he prayed, sometimes all night long. But by the time he left Europe after three years in the service, he was also proselytizing for his new faith, visiting local scientific conferences and passing out antievolution tracts, decrying the tragic predicament of the unsaved masses on the eve of Christ's return.

He decided to go back to college, only this time one that suited his current disposition—Evangel University, founded by the Assembly of God ministries. Even in that like-minded environment, Dowd's commitment and passion stood out. "I walked in to class one day and I saw that they were teaching evolution. I walked out, quite certain that Satan had a foothold in the school." He smiles wryly at the memory. "I remember saying to my roommate that I bet there are only seven *real* Christians on campus. And I was certain that I was one of the seven."

Throwing himself into studying philosophy, history, and literature with characteristic passion, Dowd discovered that his once drug-addled brain still worked quite well, and he soon proved to be one of the top students at Evangel. And he also learned that some of his fellow students and teachers—committed Christians with whom he worked, studied, and prayed—actually believed in things like evolution. The more he learned, the more his worldview began to open up.

The final step in Dowd's second conversion was a meeting with a Christian-Buddhist hospital chaplain, Tobias Meeker—a liberal-minded former Trappist monk who fully embraced evolutionary science and had a personality Dowd would not forget. Meeker was "the most Christ-like man I had ever known up to that point," he explained. "My head was saying, 'I need to save him.' My heart was saying, 'I want to be like him.'" Eventually, the heart won out and Dowd's journey as an Evolutionary had begun.

Decades later, at that lecture in Boston, I was enjoying the unique result. In a postmodern, ironic world that often seems to have conflated all deeply felt spiritual conviction with Jimmy Swaggart–style fundamentalism, Dowd is something of an anomaly. And that seems to be at least one of his underlying messages—that it's time to venture back into the waters of passion and conviction, fully supported by the open-minded curiosity of science and the inspired idealism that comes from appreciating the position in which fourteen billion years of evolution has placed human consciousness. "What evolutionary spirituality offers is a confidence, a groundedness in truth, that the liberal churches have lost," Dowd explains. "Liberal Christians so often lack the passion. They don't speak from that base of confidence. And now, with this Great Story perspective, we can all begin to speak again with that level of confidence."

When Dowd speaks about the Great Story of evolution, he is talking largely about the truths of science. No longer caught up in defending the stories of scripture, he argues for the need to move beyond what he calls "flat-earth faith," meaning faith that was forged in a cultural context uninformed by today's scientific knowledge—which includes the primary scriptures of almost all the world's religions. Dowd's ministry is largely about taking the evolutionary truths of science and communicating them to audiences in ways that are rich with spiritual meaning and religious overtones. To find God, he believes, we need look no further than the inherent creativity of the entire cosmic evolutionary process. God, in his eyes, is really just another name for the totality of reality. Some might call him a pantheist, one who interprets the divine as being synonymous with the natural world. But he points out that pantheism, as an idea, was conceived in the eighteenth century, in a world unfamiliar with the marvelous reach of our contemporary scientific worldview. That's why he prefers the term "creatheism"—a word that captures the mixture of theism and creativity that he feels better fits the universe as we know it today.

Dowd's theological rendering marries scripture with evolutionary psychology. Our untamed human proclivities to overindulge in things like sex, food, substances, and so on are explained not by original sin but by our "lizard legacy." Instead of lamenting the fall of Adam and Eve, he suggests that we recognize that these are deep human tendencies hardwired into us by evolution, instincts meant for another time that can and will undermine our lives today if we let them. We have a choice, as individuals, to become conscious of the evolutionary programming of our past and judge whether or not it will serve our growth and development in the future. "Understanding the unwanted drives within us as having served our ancestors for millions of years," he writes, "is far more empowering than imagining that we are the way we are because of inner demons, or because the world's first woman and man ate a forbidden apple a few thousand years ago."

We have that choice as individuals, but ultimately as a species as well. And the more we collectively wake up to our deep-time evolutionary legacy and stop dismissing it as inherently antireligious or antimoral or otherwise inconsistent with a meaningful life, Dowd suggests, the sooner we can start to collectively develop the kind of broad-based shared commitment to our evolutionary future that is so needed. We can begin to make the kind of choices that align our self-interest with the interests of that larger planetary process that enfolds our species. The more we wake up to this evolutionary view of reality, the more capable we are of consciously and positively directing our own development as a species.

To help further the awakening of this shared commitment, Dowd makes a distinction between private revelation and public revelation. Private revelations are those personal spiritual experiences that we may consider important but are empirically difficult to verify. Public revelations, which in Dowd's eyes have greater value, are verified by a community of inquiry. "New truths," Dowd explained, "no longer spring fully formed from the traditional founts of knowledge. Rather, they are hatched and challenged in the public arena of science." Religious people often don't fully appreciate, he laments, the "revelatory nature of science," although, he admits, most people simply have not been exposed to ways of thinking about evolution that glorify God while still embracing science.

This depriviledging of private revelation may be hard for some more spiritually inclined readers to accept. Indeed, Dowd's science-inspired faith might not satisfy those who like their religious infusions laced with a little more mystical import. His approach to evolutionary spirituality tends to focus on those forms of knowledge that are the current strengths of the sciences—astrophysics, biology, chemistry, psychology—an emphasis that can bypass the subtler interior landscapes of consciousness and culture, which have been late to the party of empirical investigation. Those areas are neither his strength nor his passion. But Dowd's point about knowledge being verified by a community of inquiry is important, and it's one of the reasons why spirituality, which today trends heavily toward a private, individualized orientation, has fallen out of fashion in the publicly minded, peer-reviewed context of the contemporary search for knowledge. But in theory, there is no reason why private revelation couldn't itself be open to public inquiry, challenge, and debate. I suspect that evolutionary spirituality in its future forms will embrace much more transparent, exploratory collective modes of inquiry, even in regard to the more intimate interior landscape of the self and soul.

Whatever the case, Dowd's inspiration is infectious and his influence has been far-reaching. He is wonderfully open-minded, loves debate and discussion, and his big-tent approach has won him the respect of leaders across the cultural spectrum. "As we integrate the Great Story of cosmogenesis, the epic of evolution, into our lives," Dowd declares, "we will see a worldwide spiritual revival." The reason is simple. "It is a story that includes all of us. In this Great Story, there is no human story that is left out."

THE MATRIARCH

The term "conscious evolution" conveys a great deal of meaning in its simple formulation. The basic idea is straightforward: now that we have become aware of the evolutionary process, conscious of this vast context that has produced human agents with at least some measure of free will, our choices matter a great deal. Today we can choose to direct our own destiny. No longer must we unconsciously stumble through this event called human life with little sense of whence we came, clinging with closed eyes to ancient myths or outdated worldviews, staggering from crisis to crisis, reacting as best we can to the news of the day. Finally, after billions of years, evolution achieved a remarkable breakthrough. It created a being that has the capacity to understand what's happening! Even then, it took many thousands of years for that species to start to grasp the nature of the process it is a part of. But little by little we have opened our eyes and started to glimpse the enormity of the picture. We should not underestimate the import of our moment in this history. After eons of blind, unconscious evolution, a creature exists who can decide to consciously evolve.

Barbara Marx Hubbard has done perhaps more than any single Evolutionary to spread the idea of conscious evolution far and wide. It is a commitment she embraces with a deep, almost maternal care for the fate and potential of our species. She has been instrumental in introducing the whole idea of futurism to our culture. Buckminster Fuller noted as much, once calling her the "best informed human now alive regarding futurism." And now in her eighties, she has lived her philosophy to the hilt, never resting upon her considerable achievements, even as she has watched the culture over the last decade begin to catch up to her prescient visions.

Unlike most of the Evolutionaries in this book, Hubbard is not a boomer. She came of age in the years when most boomers were still enjoying Gerber baby food and learning to crawl. Born Barbara Marx in New York City in 1929, she was raised in a wealthy Fifth Avenue family. Her father was a self-made entrepreneur who propelled himself up from nothing, truly living the American Dream. Her parents were Jewish, but her father was an agnostic, so Judaism was never an identity of the family. Whenever she would ask her father what religion they were, Hubbard recalls, he would tell her, "You're an American. Do your best." He passionately believed in the possibilities of the modern world.

A turning point in Hubbard's life, as in the lives of many of her contemporaries, was the end of World War II and the bombing of

Hiroshima and Nagasaki. What was all of this power for? she wondered. She had never believed in religion, but now the religion of the modern world—progress through technology—was called into question, and she was thrown into her first existential crisis.

Within a few years, she was at Bryn Mawr College, still trying to find new meaning, "trying to be an existentialist," as she puts it. She loved to read philosophy but struggled to find a philosophical home among the popular icons of the time. She read Nietzsche, but the idea of the Superman had been thoroughly corrupted and warped by the Nazis. Karl Marx, as she told me, may have seen an important truth—a society built on the principle "from each according to his ability, to each according to his need"—but that had been turned into totalitarianism. Science had its own brand of pessimism, with Bertrand Russell declaring that the heat death of the universe was unavoidable and that to believe anything else was idiotic. Even a brief flirtation with existentialism provided no lasting satisfaction. Perhaps it was because she had retained some deep kernel of her father's optimism, or maybe it was because she could brook no philosophical advance that seemed so much like a retreat, but she struggled mightily with the idea that there is no true meaning to the universe except that which she gave it—a disturbing notion that simply could not take root in the not yet fully formed structures of her young mind.

She even flirted briefly with Christianity but could not reconcile herself to the idea that Eve's guilt had caused the fall from grace. Thrown back on herself once again, this philosophically inclined daughter of America decided to seek wisdom where all budding existentialists go: Paris. And so, during her junior year, Barbara Marx boarded a plane for the city of Sartre and Camus, hoping to find meaning on the banks of the Seine.

One day, while hanging out in a café in Paris, she struck up a conversation with an artist, a man by the name of Earl Hubbard. She asked him the questions she had been asking everyone, the questions inspired by Hiroshima: "What do you think is the meaning of our new power that is good, and what do you think is your purpose?" This young man had an immediate answer: "I'm seeking a new image of man commensurate with our power to shape the future. When a culture loses its story, loses its self-image, it loses its greatness. The artist has to find a new story and until it is expressed by artists, we won't be able to bring our culture to fruition." Her only response was a small voice in her mind that said, "I'm going to marry him."

Flash-forward one decade to 1960, and Barbara and Earl Hubbard are married and living in Lakeville, Connecticut. Despite their bohemian beginnings in Paris, they embraced a relatively conventional lifestyle. She had five children ("mindless fecundity" she calls it, quoting Margaret Mead) and her inheritance supported his work and their life. But Hubbard was dying inside. She loved her children deeply, but there was something more, something deeper for her to give. Her passion for finding that "new story"—the answer to the questions that had plagued her since her teen years—was temporarily suppressed. Caught between her acceptance of the gender norms of the 1950s and the passions of her inner life, she felt like she was "turning to stone." But help was coming. The atmosphere of the 1960s was beginning to shift and Hubbard's intellectual antennae were on the lookout for anything and everything new that she could get her hands on. The first thing that triggered her curiosity was Abraham Maslow. When she read Maslow's theory of self-actualization, she had a key insight, one that framed her own depression in a completely fresh way. "I'm not neurotic," she remembers realizing, "I'm underdeveloped. I don't know what my vocation is." The difference was important; it gave a positive context, an *evolutionary* context, to her struggles and encouraged her to keep searching.

The next person to show up on Hubbard's radar was, as for so many Evolutionaries, Teilhard de Chardin, and his masterpiece, *The*

Phenomenon of Man. Only recently allowed to be published, his ideas were unconventional—and all the more so for a Connecticut house-wife. I like to picture one of the wives in the popular TV show *Mad Men* sitting down in her suburban living room after tucking the kids into bed, putting aside Betty Crocker and picking up Teilhard—it brings home how ahead of her time Hubbard was. And Teilhard's writings struck her like a lightning bolt. She found in him a kindred spirit, and a deep confirmation of her own intuition that something important was coming in our future, some new way of interpreting the human experience. "I realized in reading Teilhard that the drive inside me for greater expression, greater connectivity, and greater consciousness was the universe expressing itself as a person. Instead of being a neurotic housewife, I could see myself as an expression of a universal evolutionary process, which every single person is."

After Teilhard came Aurobindo. Then Buckminster Fuller. Soon Father Thomas Merton and Jonas Salk, and on and on and on. She was beginning to wake up to a new sense of the world and a new sense of herself. The outer conventionality of her Connecti-cut life was giving way to an inner evolutionary drive. Publishing the first evolutionary newsletter, she slowly became connected to a network of like-minded souls who had struggled with the same questions and were sensing the same need for new answers. But despite this newfound excitement, she was still just a woman in a house in Connecticut with five children, living these new visions vicariously while playing second fiddle to her husband. The illusion, however, was slowly cracking. The feminine mystique was giving way. It would take one more push for her to jump out of the nest and truly fly as a powerful evolutionary in her own right. And that push would come in the form of an evolutionary epiphany.

Anyone who has ever lived in New England knows just how cold it can get in the heart of February, when winter seems intermi-nable, summer can barely be evoked as a memory, and the prolonged chill starts to seep in around the visceral edges of consciousness. It was amid the austere cold of the Berkshire Mountains, not far from Hubbard's home, that Herman Melville once placed his stamp on the nascent American form of literature, recalling in his remarkable prose the tropical sea journeys of his youth even as the snow swirled around his house. It was this same wintery bleakness tinged with beauty that inspired Edith Wharton to settle down only a few miles from where Melville once toiled and describe in her sweeping novels the emerging politics and class problems of a rapidly industrializing America. And it was against a similar backdrop, one bitter winter afternoon in 1966 that Hubbard, a child of a now mature American power, would find the answer to her need for a new story, a new metaphor to contextualize the mushroom cloud that had awakened her conscience and shaped her life.

On that particular day, Hubbard had been reading Reinhold Niebuhr, another one-time resident of the Berkshires (though only in the more hospitable summers), and was struck by a line he had quoted from Saint Paul: "All men are members of one body." She recalls the events that followed:

> Unexpectedly, a question burst forth from the depth of my be-ing. . . . Lifting my voice to the ice white sky, I demanded to know: "What is *our* story? What in our age is comparable to the birth of Christ?"
>
> I lapsed into a day-dream state, walking without thinking around the top of the hill. Suddenly, my mind's eye penetrated the blue cocoon of earth and lifted me up into the utter blackness of outer space. A Technicolor movie turned on. I felt the earth as a living organism, heaving for breath, struggling to coordinate itself as one body. It was alive! I became a cell in that body.

Hubbard goes on to describe in great detail how this "movie" revealed to her the fundamental unity of humankind. "We are being born," she realized. "Our story is a birth of a universal humanity!"

I felt myself tumbling through an evolutionary spiral. . . . The creation of the universe, the earth, single-cell life, multi-cellular life, human life, and now us, going around the spiral once again. It all raced before my inner eyes. . . . Then, as suddenly as it had begun, the Technicolor movie of creation stopped. I found myself upon the frosty hill in Lakeville, Connecticut, alone. There was no sign of what had happened. Yet I knew it had been real. The experience was imprinted forever upon my very cells.

I had found my vocation. I was more than an advocate, I was a story teller! . . . My personal purpose was revealed to me as a vital function in the life of the planet as a whole.

More than four decades later, Hubbard's conviction in her personal purpose has only increased. She has parlayed that solitary revelation into an impressive body of work, and an even more impressive life. From the seed of the new story she had finally found, she cultivated a vision that has become her life's work: the vision of conscious evolution. Its essential task, she writes, is "to learn how to be responsible for the ethical guidance of our evolution. It is a quest to understand the processes of developmental change, to identify inherent values for the purpose of learning how to cooperate with these processes toward chosen and positive futures, both near term and long range."

"Conscious evolution" is a term that can be used in many different ways, but Hubbard captures the broad strokes of its meaning in what she calls "the three C's:"—new cosmology, new crises, and new capacities. Our new cosmology is the story that science has revealed about where our universe came from, about the extraordinary process of which we are a part. It is important, Hubbard writes, because it gives us "a new sense of identity, not as isolated individuals in a meaningless universe but rather as the universe in person" Our new crises, she explains, are those potentially catastrophic global issues we face, such as climate change. In light of evolution's trajectory, these are reframed as a "natural but dangerous stage in the birth process" of our next evolutionary stage. Hubbard points out that there have always been crises as part of the process—mass extinctions, ice ages, and so on—but never before have we had advance warning of our pending self-destruction and therefore had the opportunity to do something about it. We are shifting, she writes, from "reactive response to proactive choice."

The third piece in Hubbard's triad, new capacities, are recently developed powers such as biotechnology, nuclear power, nanotechnology, cybernetics, artificial intelligence, and artificial life. She acknowledges that in our current state of "self-centered consciousness" these are potentially hazardous, and yet, she suggests, they may be exactly what we need for the next phase of our evolution. She cautions against acting out of fear and prematurely destroying these new technologies. The task of conscious evolution is rather to "guide their capacities toward the emancipation of our evolutionary potential."

Hubbard is a visionary, but she has never hesitated to make her visions practical. In the decades that have elapsed between that epiphany in Connecticut and the writing of this book, Hubbard has journeyed down several lifetimes' worth of roads less traveled. She participated in and supported many of the transformations of the 1960s and '70s but never bought into the reactionary side of the countercultural movement, or its pessimistic tendencies. She has been a consistent advocate for the positive potential of a shared future, even amid the vigorous rejections of those who saw limits and ceilings where she saw open sky and new promise. She helped to create the Foundation for the Future, was a cofounder of the World Future Society, and started the Foundation for Conscious Evolution, bringing the perspective of an evolutionary futurism to a culture in need of new storytellers. And through a multitude of conferences, thought-leader dialogues, political engagements, films, books, citizen diplomacy, educational seminars, and even a nomination for vice president, Hubbard has started a thousand small fires—in the hope that some would burn out of control. A few did. Once it was an ordeal just to connect with a like-minded soul; now she encounters them all across the country. And since that day in 1966, her conviction has never wavered. She has raised the banner of conscious evolution for half a century, even when it was only seen by a handful of children in a living room in the Berkshires.

THE COSMIC BARD

In his classic *The Hitchhiker's Guide to the Galaxy*, science-fiction writer Douglas Adams describes a torture device known as the Total Perspective Vortex. Those unfortunate enough to be thrown into it are treated to "one momentary glimpse of the entire unimaginable infinity of creation, and somewhere in it a tiny little marker, a microscopic dot on a microscopic dot, which says, 'You are here.'" Most victims of this device die instantaneously, he explains (with the exception of the ex–Galactic President, Zaphod Beeblebrox, who was therefore proven to have an ego bigger than the entire universe), demonstrating that "if life is going to exist in a Universe of this size, then one thing it cannot afford to have is a sense of proportion."

I would both agree and disagree with this statement. There's no doubt that even a glimpse of the magnitude of our universe crushes the hubris of the human ego with the recognition of its utter insignificance. But for any aspiring Evolutionary, the opposite is also, paradoxically, true. As the conscious evolutionists will tell us, in the recognition of the vastness of the process we are a part of and our unique place in it, we can also discover the enormous *significance* of our intelligence and capacity for choice. A sense of proportion, therefore, is also a very healthy attribute, one with which we too often lose touch.

It is easy, when discussing the subject of evolution, to get caught up in the ins and outs of various biological theories, or to parse the complications and cultural implications of the religion debates, or even to become captivated by the many important philosophical issues it raises. And in so doing, it is easy to forget that evolution is about more than genes and DNA and selection and adaptation or even cultural stages and spirals. It is also about the vast ocean of the cosmos. It is about awe and amazement, humility and perspective. It is not just about our connection to the Earth; it is also about our spiritual connection to this immeasurable universe out of which we have emerged.

When I want to reflect on our spiritual connection with the universe, I turn to a man whom I like to think of as a human "total perspective vortex," albeit an extremely inspiring one—Brian Swimme. A mathematical cosmologist at the California Institute of Integral Studies, Swimme has dedicated his life to helping people grasp the literally awe-inspiring and almost ungraspable context that informs the choices humans are making at this moment in history. In his magician mind, scientific facts become enriched and dimensionalized, their simple surfaces bejeweled with meaning, their ascetic truths exploding with intimate revelation. Take, for example, his description of the evolutionary process:

> It's really simple. Here's the whole story in one line. This is the greatest discovery of the scientific enterprise: You take hydrogen gas, and you leave it alone, and it turns into rosebushes, giraffes, and humans.
>
> That's the short version. The reason I like that version is that hydrogen gas is odorless and colorless, and in the prejudice of our Western civilization, we see it as just material stuff. There's not much there. You just take hydrogen, leave it alone, and it turns into a human—that's a pretty interesting bit of information. The point is that if humans are spiritual, then hydrogen's spiritual. It's an incredible opportunity to escape the traditional

dualism—you know, spirit is up there; matter is down here. Actually, it's different. You have the matter all the way through, and so you have the spirit all the way through. So that's why I love the short version.

Henri Bergson once declared that the human mind is not designed to "think evolution." But lifted temporarily by the mind of this modern mage, we can glimpse the process behind the spell of solidity, the movement of the cosmic currents within the vast ocean of matter. And what do we learn? First of all, that none of us, in this scientifically revelatory age, are in Kansas anymore. This universe is a metaphysical mystery tour of wonder. Cosmic knowledge comes at us fast and furious these days, digitized and downloadable for mass consumption. And as it does, the universe gets bigger, stranger, closer, and yet more mysterious every day—deep-space pictures, God particles, multiple universes, string theory, eleven dimensions, parallel universes, colliding galaxies, nursery nebulae, dark energy, and much more.

And still, amid that magnificent smorgasbord, perhaps the most interesting part of the universe is that we are connected to all of it. We are made of star stuff, as Carl Sagan said, we are built from a universe that has somehow been fine-tuned for life, and cosmic evolution, whatever its many meanderings, has beaten a clear path to our door. So as we twist and turn out here on this third rock from the sun, in this spiral of a nondescript galaxy in a local cluster, and as we contemplate our place, our moment, and our time, we can, for the very first time since our brains began to wonder about themselves, draw a connection from the beginnings of this particular universe all the way to us. And perhaps, for a moment, we understand that our minds and bodies, thoughts and feelings, are not only of this time and place—they are also billions of years old. In some sense, we were, each one of us, born out of and intimately connected to that same cloud of fiery hydrogen gas 13.7 billion years ago. "Not only are we in the universe," as astrophysicist Neil deGrasse Tyson notes, "but the universe is in us."

For Swimme, that connection was made when he was a young professor at the University of Puget Sound, a passionate academic but a disillusioned one as well. Deeply concerned about the environmental crisis, Swimme finally resigned his professorship and left the West Coast looking for answers. He had one name to guide him in his search—Thomas Berry, director of the Riverdale Research Center in New York City. Berry was a Catholic monk, originally of the Passionist order, whose response to the cultural ennui and religious weariness of the twentieth century was a powerful, new, ecologically relevant vision of human existence he called the Universe Story. He had joined the Church at the age of twenty, traveling widely and studying broadly, delving deep into the religions and cultures of the world. When I saw Berry speak in 2004 in northern Vermont, a few years before his death, he was almost ninety years old, a softly glowing ember of what once must have been an incandescent star. When Swimme walked through his door in 1981, Berry was sixty-seven and at the height of his powers. He was a deeply respected world scholar and historian, with a command of several languages, including Sanskrit; a deep fluency with the multiple cultures of Asia; books on Indian philosophy and Buddhism under his belt; all to go along with his many years spent contemplating Western cultural and religious traditions.

Swimme recalls the opening conversation:

> He listened carefully as I tried to explain my misery and confusion over the destruction of the planet and what to do about it. After a long pause and without saying a word, Thomas Berry pulled a book from the thousands on his shelves. With stern visage he tossed across the table Teilhard de Chardin's great work [*The Phenomenon of Man*]. . . . My disappointment was instantaneous. This was old stuff. I had come all the way across the

continent to receive a book I had read back in my Jesuit high school? . . . Berry just smiled, and broke into easy laughter. . . . He pointed to the book he had put in my hands. "Begin with Teilhard. There's no substitute for a close reading of his work."

Berry was greatly influenced by Teilhard, but his work had also sought to modify the master in significant ways. Berry brought forth a much more ecologically oriented vision of evolution, one more in tune with post-1960s environmental sensibilities. He softened Teilhard's anthropocentrism, and critiqued his optimism and unambiguous embrace of progress. In that sense, Berry was very much his own person, and I've never really considered him, as some do, the heir to Teilhard's work. Whereas Teilhard's visionary sense of the future vibrated with a transcendent energy and inspirational power, Berry's kinder, gentler approach focused more on the rich diversity in the human and earth community and a sense of humility about the role that human beings have to play in the unfolding narrative of our time. He tried, with little success, to bring Teilhard's evolutionary context to a Deep Ecology movement that was desperately in need of such a powerful, orienting story but which generally mistrusted any thinker who put much faith in modernity's sense of optimism. Though I largely observed Berry from afar, it always seemed as if there was a touch of sadness in his work, and his thoughts often reflected something of an inner struggle between a sense of faith in the future of this evolutionary experiment and a sense of profound tragedy at the destruction humans have wrought on the Earth he dearly loved:

We find ourselves ethically destitute just when, for the first time, we are faced with ultimacy, the irreversible closing down of the earth's functioning in its major life systems. Our ethical traditions know how to deal with suicide, homicide, and even genocide; but these traditions collapse entirely when confronted with biocide, the extinction of the vulnerable life systems of the Earth, and geocide, the devastation of the Earth itself. . . . The human is at a cultural impasse. . . . Radical new cultural forms are needed.

If the transhumanists represent one extreme in the diversity of evolutionary visions in this book—the transcend-at-all-costs, techno-positive, biology-is-for-wimps, let's-engineer-the-universe sense of optimism—then Berry represents perhaps the other side of the picture, a corrective argument for a radically biocentric approach to the future, a deconstruction of human arrogance, and a deeper embrace of our immanent spiritual connection with the natural world. A future evolutionary worldview certainly needs the insights of the latter, but neither can it thrive without the forward-looking spirit of the former. The amazing thing is that Teilhard inspired them both.

Swimme, in turn, would bring his own unique qualities to the intellectual heritage of Berry and Teilhard: a scientist's sensibility and an unsurpassed talent for communicating the majesty of the Universe Story. But before he could realize his own destiny, Swimme's apprenticeship with Berry had another surprise waiting. As Swimme studied Teilhard, in the quiet of his room and in discussions with his new mentor, he began to suspect, he recalls, "that the fundamental categories of my mind were undergoing some sort of change. The unexamined assumptions that had been organizing my experiences in the world were now writhing under the pressure from Teilhard's massive and penetrating cosmology." One day, the pressure reached a climax in a powerful epiphany that brought home the living power of our connection to the deep-time context of the universe's emergence:

I watched my four year old son climb on top of a large boulder in a deciduous forest just north of New York City. The rock, staying just what it was, suddenly became molten, and my son, staying just as he was, also became molten as did the cool forest shade and the multicolored leaves—some damp, some rotting— and the dark burbling stream. All of it blazed now with the same fire that had flared forth in the beginning and was now in the form of this forest. . . . I finally understood what Teilhard was saying.

One of my colleagues once described Swimme as a nature mystic of the twenty-first century. I couldn't agree more. But it is also instructive to note here how different Swimme's experience of "nature" is from the experiences of the mystics of earlier periods in history. Whereas Wordsworth, Shelley, and the other Romantics had powerful experiences of communing with the natural world, I would suggest that the experience of "nature" in Swimme's epiphany has taken on something of a new character in this evolutionarily informed day and age. Indeed, back when Wordsworth's mind was filled with the joy of elevated thoughts as he gazed upon Cumbria's lakes and mountains, the natural world was still a stationary event, a magnificent yet intimate canvas against which the Romantics were able to reflect both on themselves and on the conditions of modernity. But the essential character of Swimme's nature is not stationary at all—quite the opposite. Things are in motion, there is a deep inherent sense of time, and the core revelation is of the vast richness of the universe's becoming.

Swimme, as he puts it, "never recovered" from his epiphany, and thus began a rich collaboration that spanned the next two decades. Together, he and Berry published *The Universe Story* in 1992, outlining a new evolutionary vision for what they dubbed a new age of the Earth, the "Ecozoic era." And Swimme's first single-author book, *The Universe Is a Green Dragon*, is based on a series of dialogues between a boy and his teacher, Thomas, who is partially modeled on Berry. It captures the personal side of Berry, the lightheartedness of the wise elder and master, weaving a tale of insight and wonder.

Swimme's calling in life is to tell stories that help situate the human experience and the human power of choice against the newly revealed backdrop of an evolutionary process in which so many remarkable events are converging all at once. For example, there is an amazing amount of new information about the nature of the universe that simply wasn't available before. "Take the discovery of cosmic evolution, the realization that the universe is expanding," Swimme proposed in a recent interview. "It's such a shock. The universe isn't just a place, *it's a movement*. And . . . we now realize it began 13.7 billion years ago. Even in the early twentieth century, we didn't know if there were two galaxies in the universe, and now we know there are at least 100 billion. There's an explosion of knowledge around evolution and the universe, and we're challenged by what it means to take this in, because we're discovering that what it means to be human is now different than before. Yes, we are individuals who are part of cultures, but at the same time, we *are a dimension of the entire universe*." Like other Evolutionaries in this book, Swimme compares the significance of our time to the Axial Age, when so many of the great religions came into being.

When Swimme remarks that "we are a dimension of the universe," it is perhaps hard to appreciate what that might signify. But I think it is reflective not only of Swimme's philosophy but of his own way of approaching the evolutionary process. It is often pointed out that one of the great developmental challenges for human beings is to genuinely be able to take the perspective of another, to stand in someone else's shoes, so to speak, and imagine life from their point of view. It is this critical capacity of empathy that allows us to actually understand the perspectives, frustrations, and sufferings of others and, ultimately, to better negotiate the multiple perspectives and worldviews that are operating in human culture today. But it strikes me that Swimme's particular gift is that he attempts to stand in the shoes of the universe itself, to place himself into the mind of the evolutionary process, to imagine the outlook from *that* point of view, to glimpse the intention behind that vast creativity and intelligence. Sure, I can appreciate that evolution came up with eyesight and that it was an immensely complex, delicate process. But I could read a thousand textbooks and never get to this:

> The Earth wants to come into a deeper way of reflecting on itself. The invention of the eye is an example. It's almost like the life process wants to deepen its awareness. It first invented eyes that were made out of calcite, a mineral. It was so desperate to see, it actually found a way to see using a mineral. Scientists estimate that life invented eyesight *forty separate times*. It wasn't an accident. It is as if the whole system of life was going to find a way to see one way or another. So what's the essence of life? Life wants a richer experience. Life wants to *see*. And we come out of this same process. We also want to see, we want to know, we want to understand deeply. That is a further development of this basic impulse in life itself.

The context for Swimme's message is a deep concern over the state of our planetary community. He points out that an entire geological era may be coming to a close in our time. We are, scientists tell us, in the midst of a mass extinction unlike anything that has happened since a meteor sent the dinosaurs scrambling for cover. "We happen to be in that moment when the worst thing that's happened to the earth in sixty-five million years is happening now," Swimme explains. "That's number one. Number two, *we* are causing it. Number three, we're not aware of it. There's only a little splinter of humanity that's aware of it."

Massive expansion of knowledge. Widespread contraction of species. And at the very moment when all of this is dawning on us, we are realizing that we have a unique role to play in the unfolding of the next stage of this planet's destiny. As Swimme points out, the force of natural selection has, in this era, in some sense been super-seded by human choice: "It's amazing to realize that every species on the planet right now is going to be shaped primarily by its inter-action with humans. . . . It is *the decisions of humans* that are going to determine the way this planet functions and looks for hundreds of millions of years in the future. We *are* the planetary dynamic at this large-scale level."

It's what we do that matters—for us and for every creature on the planet. Again, the point Swimme is bringing home is that evolution is not just happening *out there* in nature. It is happening *in here*, in the choices we are making every day. It's all about our choice.

In fact, choice itself may be foundational to evolution. If we put on our slightly speculative hats for a moment, we might observe that human free will and choice is itself an evolved form of agency, a quality that arguably can be traced back through our evolutionary past. There is no evidence for anything like agency in the crashing, spinning, merging, expanding world of the early universe, though perhaps we might catch a fleeting glimpse in the very expansion of the cosmos itself, the initiatory force that set everything into mo-tion. However, billions of years later, the universe went through a major transition, what philosopher Holmes Rolston III calls the second Big Bang, the emergence of life. What is life? It's a mystery that has been debated by scientists and philosophers for centuries, but one recent avenue of thought was put forward by complexity theorist Stuart Kauffman. In his book *Investigations*, he theorized that the very essence of life was the birth of autonomous agency. Poetically expanding on this theme, he writes: "Some wellspring of creation, lithe in the scattered sunlight of an early planet whis-pered something to the gods who whispered back, and the mystery came alive. Agency was spawned. With it, the universe changes, for some new union of matter, energy, information, and something more could reach out and manipulate the universe. . . . Agency may be coextensive with life."

Rolston proposes that there was also a "third Big Bang"—the interior explosion that gave birth to the self-reflective human mind. Agency is again fundamental to this breakthrough. Human autonomy and freedom of choice is the radical result of this leap forward. In biology, evolution itself was the choice-maker, the default selector of fitness and adaptation, the director of nature's unconscious script. But with the mind of the human, evolution becomes conscious and now rides on the back of a superempowered form of agency, the terrible and wonderful fact of our capacity to consciously choose. The evolution of consciousness over the last ten thousand years can be tracked as the ongoing development of our capacity to stand apart from our instincts and conditioning, and to make novel choices. Agency, then, has been there all along, and our future may depend upon our capacity to exercise that same agency, now evolved into the uniquely human freedom to choose.

If all of this sounds anthropocentric, so be it. Critics may complain, but from my perspective, the power of our human autonomy does make us special. It doesn't mean we are God's gift to the Earth or that we're fundamentally separate and distinct from other species. But neither are we just another creature, one among millions. Indeed, I would suggest that the problem is not simply that we are anthropocentric, it is that our anthropocentrism is insufficient and impoverished. We do not yet fully understand the nature of our position, the evolutionary dynamics of our emergence, the precarious context of our choices, the great responsibility of our power. It is the gift of Evolutionaries like those profiled in this chapter to make us aware of the nature of this power. Conscious evolution is the willingness to shoulder the responsibility it confers.

Much has been said about the significance of our moment in history. The rhetoric has been ratcheted up, often to decibels that strain credulity and play to the extremes. Some of it has focused on the apocalyptic downsides, some of it on the unprecedented upsides, and some of it, as the three individuals profiled here exemplify, has begun to express a deeper appreciation of what an evolutionary context offers to the conversation. Now, I don't claim to know exactly which marvelous possibility or disastrous scenario will unfold as we tumble forward, mostly unconsciously, into an uncertain, fast-changing future. But what I do know is this: there are no guarantees. Failure is an option that we could easily choose by default, by not embracing the challenge of our moment. Conscious evolution, after all, is a double-edged sword. It's not just that we have the *opportunity* to consciously evolve but that we *must*. We cannot rely on the blind forces of history to propel us through the crisis this time. Neither can we recuse ourselves from the responsibility that has been placed into our hands as a species. Simply put, unconscious evolution may mean no evolution at all—not to mention trouble for most other species on this planet.

The opportunities are extraordinary. The downside is frightening. The moral equation is like none we've faced before. The task is radical but simple—to find the courage to take the reins of destiny into our own hands. To exercise our evolutionarily empowered capacity for agency, assess our situation, and consciously evolve.

It is, after all, our moment of choice.

CHAPTER FOURTEEN

The Evolution of Enlightenment

I think the sages are the growing tip of the secret impulse of evolution. I think they are the leading edge of the self-transcending drive that always goes beyond what went before. I think they embody the very drive of the cosmos toward greater depth and expanding consciousness. . . . I think they disclose the face of tomorrow.

—Ken Wilber

Allow yourself to deeply take no position. Allow everything to fall away until you are resting in and as pure Emptiness." The words were timeless, transmitting the same current of stillness, peace, and inner freedom that the enlightened teachers of the East have shared for millennia. But they were about to lead somewhere that few if any of those great sages ever ventured. "Emptiness has no qualities, absolutely no qualities," the speaker continued—not a bearded, robed monk but a clean-cut Jewish American in his midthirties. "But if one is able to stay perfectly in that Emptiness, there is a quality that begins to manifest right above the surface. It has something to do with the source of all life, with Love, and with an evolutionary impulse."

The speaker was Andrew Cohen, and the year was 1991. It was the first time I ever heard the term "evolutionary impulse." Little did I know that those two words and the marriage made that evening between the timeless wisdom of enlightenment and the idea of evolution would become the foundation of Cohen's future teaching and set the direction of my own life's work. But it must have made an impression, because I always remembered that description of an impulse arising just above the surface of emptiness, like a primordial ripple—the evolutionary power of life itself, arising in our own consciousness as something comes from nothing in the depths of the self. More than a decade later, Cohen would begin to explicitly refer to this teaching as a new kind of enlightenment, one that embraced the inherent creative power at the heart of an evolving world.

Today, Cohen's name is almost synonymous with the idea of "evolutionary enlightenment"—the fruit of his efforts to bring an evolutionary worldview to the Eastern mystical traditions, to weave *being* and *becoming* into one seamless spiritual vision. His journey has been taken in public, in the pages of the magazine he founded, *EnlightenNext*, where his dialogues with spiritual leaders from every tradition, as well as with scientists, philosophers, psychologists, activists, and more have shed light on the many facets of contemporary spirituality and set a new standard for spiritual inquiry. His vision, however, is not merely philosophical. He is a teacher and mentor to hundreds if not thousands around the world, working to create a contemporary movement of Evolutionaries—a global network that is one of the first spiritual communities of practice based explicitly on an evolutionary worldview.

That night in 1991, however, while the message hinted at something new to come, the setting was decidedly rooted in the old. Cohen was leading a retreat in the small northern Indian town of

Bodhgaya, the very place where the Buddha is said to have sat under the Bodhi tree twenty-five hundred years ago and attained enlightenment. This young American teacher was making an impression in this pilgrimage town where Western seekers come to practice meditation at the local temples and busloads of Tibetans descend every winter from their mountain homes to pay homage to the great founder of Buddhism. Amid the heat, dust, barreling buses, and intolerable mosquitoes, I remember the spiritual atmosphere of the place, the quiet village evenings, the powerful chanting at the local temples, the uncommon devotion of the Tibetans as they did their prostrations, one by one, in the presence of the legendary Bodhi tree, and the early-morning calls to prayer as the Muslim minority made their presence known in this Hindu/Buddhist town.

In those days, not long after I first met him, Cohen's teachings were much more focused on enlightenment than evolution, reflective of the spiritual tradition out of which he came, Advaita Vedanta, the non-dual school of Hinduism founded by the Indian sage Shankara in the ninth century. But Cohen was not a traditionalist, and he was determined to shake up the progressive spiritual scene, which he felt had grown settled and stagnant in the years since the first wave of seekers from the West had headed off to Asia in search of mystical awakening and liberation. Cohen's teachings were refreshingly empty of rites or rituals; and he taught without the support of any extra philosophy, dogma, or traditional props. In an East-meets-West world grown accustomed to the stiff formality of Zen monks, the strange exotica of Tibetan Buddhism, and the enigmatic smiles of Eastern gurus, this young American with wisdom beyond his years was a true novelty.

In keeping with his no-nonsense New Yorker personality, Cohen's teaching style at that time could probably best be compared with the "sudden" school of spiritual enlightenment. He was convinced that one didn't need years and years of practice to realize spiritual freedom, but simply a clear, unambiguous intention and a passionate desire to go "all the way." This teaching style reflected the circumstances of his own awakening, and was deeply influenced by the method of his final mentor, H.W.L. Poonja.

Cohen had met Poonja in 1986, at the end of a long period of traveling and seeking, much of it in India, where he meditated, attended retreats, encountered different teachers, and adopted different practices, pursuing his own enlightenment with an unusual seriousness and determination. Poonja was then a little-known teacher, despite having an impeccable pedigree as a direct student of the great Indian sage Ramana Maharshi. Cohen's time with Poonja was short, but it was enough to catalyze a life-changing transformation. In their very first conversation, it became suddenly clear to this young seeker of liberation that "I had never been un-free." As he recalled years later, "In the innermost depths of my awareness I saw a brook that was moving swiftly downhill and I knew that the unrestricted freedom of that fast-moving water was my own natural state and had always been so. I knew then, without any doubt, that unenlightenment was an illusion."

Over the next several weeks, Cohen continued to undergo a dramatic awakening—one that would soon launch him into the unexpected role of teacher and guide to others on the spiritual path. The effect of Cohen's transformation on friends and acquaintances was immediate, and his ascension from seeker to teacher to leader of a movement was lightning-fast. Within a year he was traveling around Europe, teaching every night, and within only a few years, this once nondescript seeker found himself responsible for a global community—"a job I definitely had not been looking for," he later reflected.

While supporters lauded the depth of his spiritual insight and critics voiced their apprehensions about his remarkable confidence combined with relative inexperience, both missed a quality of Cohen's character that would set him apart from his contemporaries and ultimately prove more decisive to his future than any particular spiritual vision or mystical breakthrough. He had a strong desire to learn and a willingness to evolve. While his spiritual awakening had been powerful, it had come with no traditional instruction kit by which to make sense out of the extraordinary complexity of the human condition. He had no encompassing philosophy, no theology, no textual authority—just a simple message: *Realize and respond*. Realize enlightenment, and everything else will take care of itself. The power, the liberation, the freedom was in the sheer simplicity, and the transmission from teacher to student—the key was in the internal transformation, not the external instruction.

Over time, however, this bare-bones approach began to prove inadequate to the unique demands of his new role. While many of his early students experienced powerful revelations comparable to his own, they found it difficult to sustain the transformations amid the challenges, inner and outer, of life in a postmodern Western context. But Cohen didn't hunker down and defend the timeless, pristine simplicity of enlightenment against the perceived ignorance of the phenomenal world of ego and illusion. No, he did what any well-adjusted person would do. He began to ask questions, lots of questions. What is the meaning of spiritual realization today? What does it mean *now*, in our messy, confusing global village? What is its significance, not just for the individual but for human culture? What makes a person change? He had many questions—so many, in fact, that he decided to start a magazine with a question as the title: *What Is Enlightenment?* It was a sincere question that would take him almost a decade to answer.

In order to truly appreciate the answer he would come to, as well as the broader sea-change in spirituality it would reflect, I need to take a diversion from Cohen's story and travel back in time and across the ocean—back to the time of the first great wave of Evolutionaries (the early twentieth century), and down to the southern half of the Indian peninsula, where two profound and contradictory visions of the spiritual life were vying for prominence in that ancient land of enlightenment. In these two visions, we can see the broader philosophical roots of Cohen's struggle to reenvision enlightenment for an evolutionary age.

EVOLUTION IN THE LAND OF ENLIGHTENMENT

Ask anyone to name a great figure of twentieth-century India and the response is likely to be unanimous: *Gandhi*. Who could argue? Mother India was blessed to be led through the turbulent waters of the early twentieth century by one of the most impressive individuals she has ever produced. But in addition to giving birth to perhaps her most powerful social revolutionary, the last century also witnessed India give life to one of her foremost mystic sages, Ramana Maharshi, and one of her most brilliant spiritual visionaries, Aurobindo Ghose, more commonly known as Sri Aurobindo. While Cohen started out his teaching career in the lineage of the former, he would eventually find himself much closer to the sensibilities of the latter.

In these two giants we see two radically different visions of enlightenment and the spiritual path—one based upon timeless being and one based on evolutionary becoming. We see one person focused on a dimension beyond and behind the manifest world and the other fascinated by the powers and possibilities that promote the evolution of the manifest world. And we see two diverging paths that may represent a critical juncture in the history of mysticism.

For those unfamiliar with Maharshi's story, it is a classic of Indian lore. In the late nineteenth century in the South Indian region of Tamil Nadu, a common village boy underwent a powerful and unexpected spiritual awakening at the age of seventeen. This revelation completely transformed his sense of self, and he proceeded to leave home without telling a soul. He wandered off, half absorbed in an altered state of consciousness, and days later wound up in a temple at the base of the sacred South Indian mountain of Arunachala, near the town of Thiruvannamalai. Essentially removing himself from the world and society, he threw away all possessions and thoughts of his former life, and he spent a couple of decades in silent meditation in the caves and temples around the mountain and local village. Little by little, a collection of disciples gathered around him. Eventually, and somewhat reluctantly, he began to teach, becoming one of the most renowned sages in the world by the time of his death in 1950, catching the attention of the likes of Carl Jung and providing the inspiration for W. Somerset Maugham's classic *The Razor's Edge*. And he never left the foot of that mountain.

Maharshi's teachings were simple but powerful—a version of self-inquiry that encouraged the seeker to follow the question "Who am I?" until he or she realized the source of consciousness directly, realized the Self behind all manifestation, and through resting in that consciousness forever, could win final and complete liberation. He was a teacher fully in the mold of India's mystical giants, a deeply enlightened human being who represented the transcendent Vedantic religious tradition with all the authentic power of his own unmistakable awakening:

> Existence or Consciousness is the only reality. Consciousness plus waking we call waking. Consciousness plus sleep we call sleep. Consciousness plus dream, we call dream. Consciousness is the screen on which all the pictures come and go. The screen is real, the pictures are mere shadows on it.

Maharshi died more than half a century ago, but you can still feel something of his tremendous spirit vibrating in the hot and dusty plains around Arunachala. When I visited his ashram in the early 1990s, a slow but steady stream of Westerners had started to make their way to Marharshi's former home to pay tribute to his legacy, study his minimalistic teaching, and meditate beside the same mountain that once inspired the great master. Swamis rambled along on the roads, and a number of ashrams were nestled at the base of the hill. In fact, the whole mountain seemed like a monument to the enlightenment tradition of the East.

My stay there was peaceful and invigorating. I meditated a great deal, walked around the mountain, and visited some of the local holy men, several of whom had been students of Maharshi himself. One evening, as the hot Indian sun was beginning to settle on the western plains, I set off up the mountain, looking to get a front-row seat for what promised to be a beautiful sunset. After scrambling up a few hundred feet, I found a rocky outcropping that afforded a tremendous view of the expanse of land below, and from which I could observe the brilliant crimson colors splash themselves across the sky. Soon I discovered that I was not the only person to be so inspired. A young Indian sadhu (spiritual renunciate), dressed in orange robes, soon joined me on the outcropping and introduced himself, and we sat together watching the sunset, talking about the spiritual life. A bright young man; he was, like me, in his early twenties, and he had given up all worldly ambitions to seek God in the style native to his homeland. He talked to me about his hopes and dreams and we shared our respective thoughts, passions, and plans for the future.

We must have talked for the better part of an hour on that ledge. It was a kind of magical encounter, one of those rare moments that can only really happen when you're young, on the road, and have time to spare for exactly those kind of chance, unexpected meetings. As darkness descended, I eventually made my way back down the mountain toward my sleeping quarters in a local ashram. I was left with the impressions of a deep conversation, and I felt like I had a lot in common with this young Indian, despite the fact that we came from vastly different backgrounds and would likely lead divergent lives. I was entirely rooted in the modern world, and looking to embrace a spiritual life that led me forward into the future—whatever that might mean. My young Indian friend had embraced an ancient path, one that was rich and beautiful, full of tradition and all the dignity that goes with it, but one that seemed as far from the modern world as the lives of the saints who had meditated for centuries on these hillsides. Perhaps it would lead him to happiness, I thought, or better yet, to genuine enlightenment, but it was simply not a direction I could follow.

The overall impression of the almost archetypal nature of that meeting has only deepened with time. It seemed as if I was meeting myself on that mountainside, the same young man in two different periods of history. Chance intervened, our paths met, and two roads diverged—one into an uncertain future, the other into a timeless past.

Spirituality today faces a similar choice between two distinct paths. It can resign itself to the well-worn grooves of its former glories—its mythic power, moral insights, and mystical achievements. Or it can venture forth into virgin territory, discovering a new role for spirit in the consciousness and culture of tomorrow. There is nothing inherently wrong with continuing to mine the depths of tradition's impressive accomplishments. But if spirituality wants to be more than a respite from the world, more than an alternative to the cares and conundrums of life as it marches by at a distance, it must change. If it wants to reclaim its old power as maker of history, not just witness; as contributor to the future, not just an escape from its burdens; then it must discover a new relationship between the insights that lie beyond time and the world that marches forward *in* time. "I hazard the prophecy," Alfred North Whitehead once wrote, "that that religion will conquer which can render clear to popular understanding some eternal greatness incarnate in the passage of temporal fact."

The timeless power of enlightenment has rarely shone as brightly into this world as it did through the luminous gaze of Ramana Maharshi. And yet no hints of any new religion were to be found in the red dirt pathways at the foot of Arunachala. To catch a glimpse of that brave new future we would need to venture northeast, to the edge of the Bay of Bengal, where Sri Aurobindo once sought refuge from the British colonialists in the oceanside town of Pondicherry.

It is some kind of irony of history that two of the greatest sages of the twentieth century, each modeling a completely different vision of the spiritual life, would end up living so close to each other during the exact same time period. Sri Aurobindo never met Ramana Maharshi, though the two teachers lived only a few hours' bus ride apart. Ramana never left his beloved mountain and Aurobindo never left Pondicherry, the French colony that offered him political asylum and spiritual haven for the last four decades of his life.

Aurobindo's upbringing could not have been more different from Ramana's. Born to a Brahmin family, he was sent to school in London at the age of seven and ended up at Cambridge, one of the top students in his class. After university, he headed back to India and began to familiarize himself with the culture of his homeland. Change was afoot in the land of his birth, and it wasn't long before this bright young Indian inserted himself right into the middle of the independence movement. A natural orator with a sharp tongue and sharper intellect, Aurobindo rose rapidly in the movement, eventually becoming its political leader, decades before Gandhi would assume that role. Once referred to as "the most dangerous man in India" by his British overlords, Aurobindo was focused on political revolution rather than spiritual evolution. His religious career didn't even begin until a fateful encounter, at the age of thirty-four, with a yogi. This simple holy man instructed Aurobindo to reject all thoughts that tried to enter his mind. If it sounds like an easy task, I would suggest that you haven't tried it. But Aurobindo took to it like a yogic savant. Within moments, his mind "became silent as a windless air on a high mountain summit." He describes how "I saw one thought and then another coming in a concrete way from outside; I flung them away before they could enter and take hold of the brain and in three days I was free."

In a few short days Aurobindo had achieved a goal that many pursue in vain for a lifetime. But ironically, to him, it was unwelcome—"precisely the experience he did not want from yoga" as his biographer, Peter Heehs, describes it. Suddenly, this political revolutionary and social activist, who deeply cared about the world and, in particular, the fate of his homeland and countrymen, was immersed in an experience that seemed to be telling him that the world was, in fact, unreal.

> It threw me suddenly into a condition above and without thought [in which] there was no ego, no real world . . . no One or many even, only just absolutely That, featureless, relationless, sheer, indescribable, unthinkable, absolute, yet supremely real and solely real . . . what it brought was an inexpressible Peace, a stupendous silence, an infinity of release and freedom.

Even given his newfound and unsought immersion in this nirvanic state of consciousness, Aurobindo managed to continue his revolutionary activities. But change was coming. Implicated in a misdirected assassination attempt that had been organized by his younger brother, he went to trial and was jailed for a year by the British government. It was in the confines of prison that Aurobindo's interior life began to expand and deepen. Meditating in his cell, he began to realize that the perception of the world as illusory was only the first step on his path, and that this initial realization was just the beginning of a much richer and more all-encompassing sense of the spiritual life. Under the watchful eye of his British jailors, Aurobindo stepped out of the tradition of his Indian forefathers and began to develop a new, evolutionary view of the spiritual life. "Nirvana in my liberated consciousness turned out to be the beginning of my realization," he wrote, "a first step towards the complete thing, not the sole true attainment possible or even a culminating finale. . . . And then it slowly grew into something not less but greater than its first self."

A year later, Aurobindo was out of prison and seemed to be resuming his political activities, but his spiritual sensibilities would not be denied. And when he heard that the British forces were closing in for another arrest, he escaped to Pondicherry, seeking political asylum. He would not leave for the rest of his life.

Safe from the attention of the British, and removed from his active role in the independence movement, Aurobindo's focus turned fully to his inner endeavors. A small band of acolytes gathered around him and he began to teach and write, publishing a magazine called *Arya*, which would become the vehicle for his philosophy. The purpose of the magazine, as he explained in the inaugural issue, was to "feel out for the thought of the future, to help in shaping its foundations and to link it to the best and most vital thought of the past." From 1914 to 1921, he published a series of articles and essays that would form the backbone of his philosophy and later become the basis for his books *The Life Divine* and *The Synthesis of Yoga*.

It quickly became clear just how much evolution would play a role in Aurobindo's unique spiritual vision:

> The animal is a living laboratory in which Nature has, it is said, worked out man. Man himself may well be a thinking and living laboratory in whom and with whose conscious co-operation she wills to work out the superman, the god. Or shall we not say, rather, to manifest God? For if evolution is the progressive manifestation by Nature of that which slept or worked in her, involved, it is also the overt realization of that which she secretly is. . . . If it be true that Spirit is involved in Matter and apparent Nature is secret God, then the manifestation of the divine in himself and the realization of God within and without are the highest and most legitimate aim possible to man upon earth.

In the same way that Teilhard was working to reenvision Catholicism in the context of an evolutionary cosmos, Aurobindo worked to reinvent Hinduism, to evolutionize its central tenets. His was a vision in which human beings, as currently constructed, are just one link in a chain that will extend far beyond our current existence into future transformations, both spiritual and physical. In a similar fashion to Teilhard, he saw evolution as progressing from matter to life to mind and then projected it forward into future spiritual or supra-mental planes and states of consciousness. Like his Jesuit counterpart, Aurobindo also mapped the connection between physical complexity and interior consciousness, noting that "the better organized the form, the more it is capable of housing . . . and more developed . . . consciousness." Less connected to the scientific story than Teilhard, he saw physical evolution as secondary to the spiritual evolution unfolding through physical forms. More significantly, he was one of the first to see the spiritual awakening of the individual as integral to the larger evolutionary progress, stating that the liberation of the individual is the "primary divine necessity and the pivot on which all else turns." In this respect, he embraced the great liberation teachings of the East. "By attaining to the unborn beyond all becoming, we are liberated," he wrote, his words echoing centuries of traditional wisdom. But then, as always, he would soon come around to add an evolutionary dimension to his message. We cannot stop with the realization of that which lies beyond becoming, he explained; rather, it is "by accepting the Becoming freely as Divine" that "we invade mortality with the immortal beatitude and become luminous centres of its self-conscious expression in humanity."

Aurobindo writes powerfully and cogently about philosophical issues that still trouble us today, while also capturing subtleties of the spiritual life that students of his work are still trying to fully grasp and put into practice. He called his path Integral Yoga to express the all-encompassing, world-embracing nature of the spiritual life that he called upon his students to live. A voracious reader and exemplary student, Aurobindo would certainly have been well schooled in Darwin's theories, but also was no doubt exposed to the grand evolutionary systems of the German idealists, whose work his philosophy certainly calls to mind. He denied any explicit influence, but some seems likely nonetheless.

Evolutionary pioneers love to have a go at naming the next stage of human evolution—the step after *Homo sapiens sapiens*. They must figure that, sooner or later, someone's attempt is going to stick. Teilhard used the term *Homo progressivus*. Author John White prefers *Homo noeticus*. Barbara Marx Hubbard likes *Homo universalis*. Biotechnology entrepreneur Juan Enríquez suggests *Homo evolutis*. Buckminster Fuller came up with *Heterotechno sapiens*. Science writer Chip Walter coined the term *Cyber sapiens*. Aurobindo named his future human the *gnostic being* and claimed that such a being would have transcended conventional ego and achieved a sort of universalized self—a self in whom the very processes of the universe (or multiverse) were internalized. A gnostic being would still be an individual but would, as Aurobindo writes, "know the cosmic forces and their movement and their significance as part of himself."

According to the sage, the gnostic being would seek the larger transformation of human culture. "This calls for the appearance not only of isolated evolved individuals," he wrote, "but of many gnostic individuals forming a new kind of beings and a new common life superior to the present individual and common existence." To this end, he formed a community around him that would attempt to actualize his vision. This initial ashram survived him and was led through the 1950s and '60s by his longtime collaborator and confidante, the Mother (aka Mirra Richard), a French painter, musician, and accomplished spiritualist in whom Aurobindo had recognized unusual spiritual capacity. In Aurobindo's work we can see an important milestone in the history of this emerging new philosophy. His was one of the first attempts to *apply* the spiritual principles of an evolutionary worldview, not just in the articulation of an inspiring vision or a beautiful philosophy but as a practical context for living, as a path of individual and collective transformation.

As the 1960s came and went, as the East came to the West and vice versa in a cultural exchange that would alter both regions permanently, Aurobindo's name was not prominent among the many teachers and teachings celebrated in the counterculture. And yet his influence was nonetheless profound. His writings had a significant influence upon a young philosopher named Ken Wilber. His student Haridus Chaudri founded the California Institute of Integral Studies and helped sponsor the visits of a number of Eastern luminaries in the 1960s. But perhaps Aurobindo's greatest influence on the West was through the life of a young Stanford student named Michael Murphy, who read his writings and was inspired enough to make the trip to India in the 1950s to visit Aurobindo's ashram, just a few years after the master's death. Murphy's experience would prove formative, and when he arrived back in the States, this young Evolutionary used his family property to open an institute based around the idea of an exploration of new human capacities. The Esalen Institute, as it was known, would be the launching point for the human-potential movement that would have a huge effect on American culture in the last decades of the twentieth century.

Aurobindo was never an influence on Andrew Cohen in his early years as a seeker and teacher. But the new spiritual current he had worked to set in motion was no doubt making its momentum felt. With the benefit of hindsight, we might look back and say that in the early '90s, Cohen's own path was beginning to leave behind the traditional enlightenment of the great Ramana Maharshi and embrace one more associated with the new kind of evolutionary vision spoken of by Aurobindo, but such conclusions are more after-the-fact interpretations than conscious calculations. Cohen knew almost nothing of Aurobindo's work at the time. Clearly, however, these two very different teachers, separated by time and geography, were inspired by the same source, lit up by that impulse that appears just above the surface of emptiness for those with the eyes and the heart to recognize it. Aurobindo, when he discovered it in his prison cell, called it the Active Brahman. Cohen, starting that winter in Bodhgaya, would call it the evolutionary impulse.

AWAKENING THE EVOLUTIONARY IMPULSE

While Cohen knew nothing of Aurobindo in those formative years, what he did know was that his own understanding of the purpose and significance of enlightenment was developing rapidly. Enlightenment, he was coming to understand, could not be contained within the private confines of the self. Increasingly, he felt that its meaning extended beyond the experience of a higher state of consciousness, however profound, or the personal transformation of a single, individual life, however dramatic. Spiritual transformation, in this day and age, must inevitably have real implications for our shared moral, philosophical, social, and practical lives—not just for the self but for society. Intuitively, he had always felt that spiritual awakening was somehow connected to the broader evolution of the human race, but giving voice to that emerging vision—giving it philosophical weight, moral clarity, and intellectual coherence—would take some time.

Always curious to engage with other spiritual leaders, Eastern and Western, Cohen made it a point to pursue dialogues and discussions with the local luminaries in whatever cities or countries he happened to be traveling through. Many of these dialogues would end up as content for *What Is Enlightenment?* magazine (later to become *EnlightenNext*), which was expanding and fast-becoming a nationally recognized forum for exploring critical questions of the spiritual life. But the magazine was more than a public forum; it was also the open-ended canvas of a living personal inquiry, as Cohen and his team of editors, of which I would become a part in the late

'90s, sought to understand how our own spiritual passions and intuitions fit into the larger scheme of a postmodern culture.

As the decade passed, Cohen continued to speak, teach, and travel, expanding his international network of students and establishing centers around the world. Always at his best in the teaching role, he was a tireless voice of spiritual optimism in a progressive spiritual culture that had grown a little bit sleepy and more than a little bourgeois. "Everybody wants to get enlightened," he would often say, "but nobody wants to change." He sensed that a deeper source of motivation was needed, and he began to speak about a different kind of enlightenment that was not for the sake of the individual but "for the sake of the whole."

The millennium came and went, and several critical lines of influence began to converge. First, Cohen and the editorial staff came across the work of Aurobindo and Teilhard de Chardin, along with contemporary evolutionary thinkers such as Brian Swimme—all in the period of about a year. A healthy dose of evolutionary perspective was transmitted to everyone in the offices of *What Is Enlightenment?* and for Cohen, the effect was particularly powerful. "I'd never really come across anything like this before," he reflects. "In Teilhard and Aurobindo, I started to hear echoes of my own passion—a passion for awakening to the truth of who we are, and then daring to allow ourselves to experience the urgency to make it manifest *in* this world with all of our being." It helped him to further a process that had already begun, to recontextualize and reinterpret his own enlightenment in an explicitly evolutionary context.

The other significant piece to the puzzle was Cohen's budding friendship with Ken Wilber. Wilber was on the front lines of the intellectual debates of the day, trying to influence the evolution of knowledge, and Cohen was on the front lines of individual and collective development, trying to get his students to actualize the shared spiritual potential of conscious evolution. In many respects, theirs was a friendship born of mutual recognition and support. Long a seeker himself, Wilber had deep respect for the enlightenment traditions of the East, recognizing in this spiritual innovator "a fresh and profound approach to spirituality grounded in his own awakened awareness." Cohen found in Wilber's work a synthetic and brilliant intellectual framework that helped to contextualize his increasingly independent spiritual sensibilities.

Teilhard. Swimme. Aurobindo. Wilber. In the spring of 2002, *What Is Enlightenment?* published a landmark issue with the headline "The Future of God: Evolution and Enlightenment in the 21st Century," presenting the converging currents of what we were recognizing to be a coherent new movement, a dawning "evolutionary spirituality" unique to our time in history. In this emerging movement, Cohen's voice was joined by others who were similarly inspired—great thinkers informed by evolutionary philosophy, brilliant theologians exploring the changing face of God, and powerful leaders inspired by the cosmic insights of science. But Cohen remained first and foremost a spiritual teacher, someone who was combining enlightenment and evolution in a way not seen since Aurobindo had breathed the ocean air of southern India more than half a century ago. Soon, Cohen began to call his teaching *evolutionary enlightenment*. In his 2011 book of that title, he explains the essence of what had driven him in his quest to forge this new spiritual path: "I believe that those of us in the twenty-first century at the leading edge of consciousness and culture urgently need a mystical spirituality and a source of soul liberation that points us not beyond time but toward the future that we need to create," he writes. "I believe the spiritual impulse today is calling us not away from the world but toward that big next step we need to take in our world. That next step will not emerge by itself—it must be consciously created by human beings who have awakened to the same impulse that is driving the process. Awakening to that energy and intelligence . . . is the source of the new enlightenment."

Since he first began to speak about that impulse decades ago in the birthplace of Buddhism, Cohen has come to more deeply understand its nature—his spiritual intuition now informed by his own study of evolutionary principles. The evolutionary impulse, he writes, is "the energy and intelligence that burst out of nothing, the driving impetus behind the evolutionary process, from the big bang to the emerging edge of the future." Cohen is not the first to posit an energetic drive at the core of the evolutionary process (Henri Bergson's *élan vital* is just one of several similar ideas), but he may be the first to specifically identify its expression at multiple levels of the human experience. He explains that the evolutionary impulse, expressed at a biological level, is felt as the sexual urge to procreate, while at a mental level, it is experienced as the uniquely human desire to innovate and create. And most important, that same impulse, he believes, is also felt spiritually as the mysterious longing to transcend self-limitations and evolve at the level of consciousness—the longing that had been driving him for decades.

Making this critical connection between the personally felt spiritual impulse and the cosmic evolutionary impulse has allowed Cohen to find a new answer to the question "What is enlightenment?" While always honoring the traditional mysticism that triggered his own awakening, Cohen feels that liberation, to be relevant in our time, can no longer merely be about *freedom from* the world and all its complexity. Rather, spiritual freedom—liberation from the petty concerns of ego, narcissism, fear, and desire—gives us the *freedom to* participate in the larger creative process of the evolving cosmos. And that freedom is not just found through letting go into the timeless depths of being, but also through allowing one's very self to be overtaken and enlightened by the energetic power of the evolutionary impulse. "Free, in this sense, means available," he writes. "Available means we are no longer endlessly distracted by the karmic momentum of the past, by the fears and desires of the personal ego or the culturally conditioned self. Only when buoyed by a measure of inner freedom from that momentum will we be spiritually awake here and now and therefore available for the overwhelming task of consciously creating the future."

While many people are inspired by the notion of conscious evolution—awestruck by its vastness, energized by its promise, motivated by its moral implications—Cohen has particular insight into what it actually takes for an individual to follow through on that inspiration. It is wisdom that has been hard won through years in the evolutionary trenches, working directly with individuals and groups. He knows from experience that no matter how inspired we are, as long as our attention is distracted by psychological issues or culturally conditioned biases and assumptions, under pressure it will be almost impossible for us to stay in touch with that evolutionary vision, let alone contribute to its unfolding. And so he has created a spiritual path and practice to address the human challenge of becoming "fit vehicles" for an evolutionary life.

The transcendent freedom of enlightenment, the creative energy of evolution—brought together into one singular spiritual vision. It is a teaching that produces neither saints nor monks nor ascetics nor mystics but rather a particular kind of Evolutionary—awake to the depths of spirit but alive with the promise of the future, and free enough to respond to a world in need of evolution. Aurobindo had sown the seeds, Cohen was beginning to reap the harvest, but in truth, I believe it was far beyond either of them. Both were carried by a larger current: the evolution of spirit itself unfolding in the depths of individual realization over time.

An evolutionary worldview, as I mentioned in chapter 2, carries with it the realization that even our deepest intuitions of spirit are not static, fixed, or unchanging, but they too are developing as history itself moves forward. Yes, there may still be animistic beliefs and shamanistic trances, totems and taboos, meditating monks, whirling dervishes, and humble souls reaching toward heaven on bent knees. There may be deists and theists and nature mystics, pantheistic passions, pagan rituals, channeled spirits, and attentive mitzvahs. There may be God-bearing witnesses to the glory of heaven above and ten-thousand-seat stadiums delivering souls to a happier home. All of these are part of the rich tapestry and history of the religious calling, many still serving important roles in the complex story of human consciousness and evolution. But at the tip of spirit's arrow, where evolution is restless and ever seeks to transcend itself, new forms and new expressions are being created, and it is here that a new enlightenment tradition is forming, a path of transformation that can liberate our spirits and strengthen our souls for the enormous tasks ahead.

CHAPTER FIFTEEN

An Evolving God

Has creation a final goal? And if so, why was it not reached at once? Why was the consummation not realized from the beginning? To these questions there is but one answer: Because God is Life, and not merely Being.

—*Friedrich Schelling*, Philosophical Inquiries into the Nature of Human Freedom

Let's face it, theology just isn't that cool. From the New Atheists to the "spiritual but not religious" to the struggles of the Catholic Church to the emptying of the old mainline Protestant churches, evidence abounds that theism in the twenty-first century isn't exactly avant-garde. Now, I'm sure it had its heyday once upon a time. Back when the Huns were ransacking the Roman Empire, theism was the very expression of progressive culture. But it's been a long time since it was quite so hip to be holy. Sure, the megachurch world of Rick Warren and Joel Osteen may have created subcultures in which it's cool to believe in the Abrahamic God, but that's different. That's more about emotional convictions than intellectual conclusions. I'm talking about theism as an "ism"—an idea, a theology, a God-centered worldview, a philosophically coherent structure of beliefs about the way the universe works that is built around a transcendent creator. And that kind of theism isn't exactly lighting the intellectual universe on fire these days.

Mention cognitive science, evolutionary psychology, or neuroscience and it's easy to find plenty of *New York Times* bestsellers examining the fascinating edges of these growing fields. Theology? Not so much. In fact, you might even imagine that nothing particularly significant has happened in theology in decades, if not centuries.

But you would be wrong.

There is, in fact, a cutting edge to contemporary theology, and that was the reason I recently found myself about thirty miles east of L.A., at the foot of the San Gabriel Mountains, looking for the Claremont School of Theology. It was here, I had learned, on the tectonic border of the nearby Sierra Madre Fault and the massive San Andreas Fault, that a new form of evolutionary theology is being formed. On the active and unstable intellectual fault line between the worlds of science and religion, a new kind of common ground was being sought.

At first glance, Claremont's small, nondescript campus doesn't exactly shout "new and different." There are a number of simple, architecturally average buildings and offices surrounding an impressive-looking chapel that stands out right in the middle of campus. I arrived a few minutes early for my scheduled interview and headed straight for the chapel, curious to get a look at the simple but elegant place of prayer. As I stepped inside and took in the stained-glass windows, the large cross hanging on one side, the simple rows of chairs facing the lectern, the hymnals at the entrance, I felt a hint of Protestant familiarity. It reminded me of my own Presbyterian upbringing, of Sunday school mornings, church socials, and confirmation. The memories are pleasant enough, but my religious upbringing was hardly formative. My family, like the families of many of our contemporaries, was not that serious about the traditional Christian God, and I remember our church much more as a community connector—as a healthy and integral part of the social fabric of a small town, not really a source of religious revelation and even less a source of intellectual engagement. It was very conventional—ordinary and slightly unremarkable in the way that I think only Protestants have truly perfected, as if the mainstream itself had a mainstream. This campus had a bit of that air about it; and sure enough, I found out later, it was founded by Methodists—not exactly God's most colorful children.

I had come here to meet Philip Clayton, a professor who was making a name for himself as a powerful thinker in the science and religion debates and the nascent field of evolutionary theology. Clayton was affiliated with the Center for Process Studies, a research center connected with the Claremont School of Theology and Claremont Graduate University. And here's where the story gets interesting. The Center for Process Studies is unique in the context of Christian theology: it is the only place in North American academic life that is specifically dedicated to carrying on the thought and work of Alfred North Whitehead and Charles Hartshorne.

Whitehead was a key figure in the first wave of Evolutionaries—those remarkable men and women in the early twentieth century who were trying to find a place for creativity, subjectivity, spirit, and even God in the context of a scientifically revealed, evolutionary universe. Initially a mathematician, he worked closely with Bertrand Russell producing the tome *Mathematica Principia* in 1900, a thousand-page thesis on the foundations of that field. Russell and Whitehead stayed close friends for life, but their philosophies eventually diverged. Russell, the great analytic philosopher, was ever the careful pragmatist; his philosophy was one of restraint, logic, and careful reason. He steered well clear of what he called the "fog of metaphysics," as well as more speculative philosophies, the latter being the area where Whitehead would make his mark.

Whitehead left behind his native England in the 1920s and took a post at Harvard in the philosophy department, the same post that had once been held by William James in the 1900s, when Harvard philosophy was a hotbed of evolutionary thinking. Whitehead's early work in mathematics would morph into a deep inquiry into the foundations of reality, eventually producing a unique body of work called process philosophy, which directly countered the increasingly dominant materialism of his day. Whereas Russell achieved great fame during his career (partially for his politics) and his immediate influence on philosophy was profound, Whitehead's work remained respected but relatively obscure during his lifetime and for decades hence. But that appears to be changing. Interest has grown significantly over recent years, and like many of his contemporary Evolutionaries, he may well prove to be a voice that speaks more to the concerns of the twenty-first century than the twentieth. We might say that while Russell embodied the mood of a war-weary twentieth century seeking restraint, retreat, and caution in its intellectual outlook, Whitehead embodied the bolder spirit of our own time, where integration, coherence, and meaning hold new prominence.

"Bertie thinks I'm muddleheaded, but I think he's simple-minded." That is how Whitehead once described the philosophical relationship between the two friends. Indeed, Russell's philosophy hewed closely to the currents of the day. He was a logician whose work helped move philosophy toward the careful and considered logic of the sciences. He was the antidote to Hegel and the grand, all-encompassing assertions of the German idealists. Whitehead's work, on the other hand, was concerned with big ideas—he was exploring fundamental connections between science, philosophy, and religion. In some sense, he followed Bergson and many of the other great evolutionary thinkers of the nineteenth and early twentieth century in carving out a place deep in the foundations of reality for change, process, movement, and creativity. He shattered the spell of solidity, seeing a universe of *becoming* where others saw only matter and stasis. He was one of the first great philosophers to fully digest the changing landscape of a post-Newtonian world and he created what some have called an ontology of becoming. Often compared to Heraclitus, the Greek philosopher who emphasized the ever-changing nature of reality and declared that a person "cannot step in the same river twice," Whitehead suggested that reality is an unfolding series of relationships between experiences in the "flux of becoming" rather than a product of the interactions of particles. Like Heraclitus's river, reality is in motion—always moving, changing, and becoming. In our ignorance, we treat the river as if it were frozen in time, and we completely miss the most critical feature that defines reality as we know it—*movement forward*—what Whitehead called "creative advance into novelty." His work encouraged us to see through the illusion that tricks our senses into attributing a kind of static permanence to the material world. He wanted to dispel the materialist haze—"the fallacy of misplaced concreteness"—and add temporality and change to our philosophy. For him, the most basic unit of reality was not bits of matter or physical particles or units of energy but rather "occasions" of experience—moments of subjective existence that fall and flow into one another in a cascade of creative becoming.

One of the reasons for Whitehead's relative obscurity in today's pantheon of twentieth-century thinkers is probably the sheer complexity of his work. Even professional philosophers struggle with his writings. But we should not let that deter us from seeking to understand his critical contribution to moving philosophy a little bit closer to an evolutionary worldview and inspiring generations to follow and expand his line of thought.

One person who did appreciate Whitehead's philosophy was Charles Hartshorne. From his posts at Harvard, the University of Chicago, Emory University, and the University of Texas, he carried on Whitehead's work, turning process philosophy into process theology, and more fully incorporating God into the picture of an evolving universe. Hartshorne studied with Whitehead for years at Harvard before heading to Chicago, where he broke with his theological contemporaries in attempting to understand God not as a complete and perfect being outside the universe but rather as a deity that was, in some sense, incomplete; a God who was becoming more perfected in the very process of the universe's becoming. With this new vision of divinity, Hartshorne rejected the ancient vision of omnipotence so common to the traditional understanding of God. He put forth a God who is actually developing as the universe itself moves forward in time. In this sense, process theology would suggest that we all participate to some degree in the being and becoming of God, in the very evolution of divinity. We are part of God's self, so to speak, and as we participate in the development of this world and this universe, so too do we, in some fundamental way, participate in God's self-development. Paradoxically, by placing limits on God's perfection, Hartshorne and Whitehead simultaneously expanded the depth of his or her being. They opened the door to seeing God not simply as an object of distant worship but as an intimate subject in whose ongoing creative self-development we can each participate.

Hartshorne's book *Omnipotence and Other Theological Mistakes* made this break with the past clear, and it was a work that certainly succeeded in riling up his fellow philosophers of spirit. Some have less sympathetically referred to Whitehead and Hartshorne's God as "God the semi-competent." But such dismissals miss the important breakthrough that their work represented in the effort to drag theology into the modern world. By drawing powerful connections between the evolutionary dynamics of the universe and the very being of the divine, they helped set the stage for a new evolutionary theology to emerge in our own time, one whose picture of divinity was at least congruent with a scientifically revealed universe. In other words, if people in this day and age are going to believe in God, then they need a God that is believable.

Between the lineages started by Teilhard and Whitehead, there is a surprisingly robust evolutionary tradition living within the walls of the Christian church—more so than in any other major religion.

In this chapter I want to explore some aspects of this evolutionary theology and bring to light how the idea of emergence, an idea that comes to us from science, is playing a role in its formation. Whether we call ourselves Christians or not, whether we find our spiritual home in devotion to a higher power or shy away from all such leanings, I'd like to encourage readers to consider that there may be important insights for an evolutionary worldview that are being forged in the rekindled theological furnaces of faith.

FROM EVANGELICAL TO EVOLUTIONARY

"Can you believe that of the people who call themselves Christian, fifty-six percent don't believe in evolution?! That's a sign of what we face in terms of denial." Philip Clayton and I had just sat down in his office at Claremont. Dressed in casual clothes and flushed from riding a bike, Clayton was a down-to-earth kind of guy, a professor who had a natural air about him and an obvious ease with people. Here was a teacher, I thought to myself, who must be well liked by his students. By the end of our meeting, I had reason to qualify that supposition, because the other quality that soon captured my attention was Clayton's rigorous intellectual intensity. Though he spoke in an open, engaging manner, it was also with a seriousness that conveyed the sense that our thoughts matter deeply, that they carry real consequence.

Clayton's long journey toward an evolutionary worldview, he explained to me, had prepared him to play a unique role in bridging science and spirit. He had been raised a passionate atheist in Northern California by parents who were professors at one of the most cutting-edge, experimental colleges of the 1960s and '70s, Sonoma State University. Activists and intellectuals, his parents played a part in the cultural and political movements of the day and their varied interests provided a rich atmosphere for their son's development.

But when the time came to assert his teenage independence, Clayton faced the same dilemma that so many children of progressive parents would face in the decades to come—there was nothing to rebel against. No declaration of sexual proclivity, drug use, or counterculture loyalties was likely to cause much consternation under the Clayton roof. In retrospect, he realizes that he chose the one thing that could make an impact. He came home one day, sat his parents down, and announced, "I've accepted Jesus Christ as my sole Lord and Savior."

It worked. As the silent tears flowed down his mother's face, the young Clayton explained how the biblical stories he had learned while at a Christian camp with some local friends—marvelous stories about Christ, Satan, Adam and Eve, redemption, and the Second Coming—had fired up his rich fourteen-year-old imagination and convinced him to commit to Jesus. He resolved to be a minister, and after graduating at the top of his high school class, he headed off to Westmont, an evangelical college in Santa Barbara.

But something critical happened to this extremely bright but spiritually narrow collegiate kid on the way to his Christian destiny. He had his second spiritual epiphany. And this one would lead in the opposite direction to the first. He had chosen philosophy as his major, convinced that it would help him preach more organized sermons. One day, he recalled, "we were in class with our favorite professor, Stanley Obitts. He was launching into this discussion of Leibniz and God, and he had a way of leaning forward in his chair and moving his hand around in the air when he spoke. It was an intense discussion, and every person in class was leaning forward as well, and then suddenly he silenced us with a stare. After a long pregnant pause, he said a four-word sentence that affected the rest of my life. He said, '*These* are the questions!' At that moment, I had an enlightenment experience. I got it. It's not about the answers. It's about the questions. It took a decade for it to fully sink in, but that was the moment I became an evolutionist."

Within a few months, Clayton had turned his attention to science and was writing about the relationship between science, philosophy, and theology. By the time he graduated a year later in 1978, his conservative evangelical faith was "in shambles" and his plans for the ministry in trouble. Something in him had changed irrevocably, but he was still struggling to come to terms with his loss of theological certainty. Heading to graduate school at Fuller Theological Seminary, he took numerous Bible classes, hoping to salvage the remnants of his shattered faith.

"Daniel Dennett calls Darwinism the universal acid," Clayton explained to me, reflecting on the way in which his embrace of an open-ended, open-minded inquiry had little by little undermined his religious conservatism. "And once the acid of questions begins interacting with dogmatic beliefs, you can't stop the acid; it eats through everything."

Clayton was fast becoming a skeptic in the land of the believers, challenging his professors to explain their biblical interpretations and justify their claims. He read more about science, and the relationship between science and religion, and found himself enthralled by the connections and disconnections between the two. But despite his unsettled heart and uncertain mind, his academic success continued. He earned a master's in religion from Fuller, and resolved to find the most interesting mind he possibly could in the field of science and religion with whom to study.

The person he chose was Wolfhart Pannenberg, a German theologian, considered to be one of the great religious scholars of the era. Like Clayton, Pannenberg was a convert to the faith—raised outside of any tradition, a spiritual experience of sacred light had changed his life in his teenage years and he became a Lutheran in his twenties. Though inspired by the Christian theologians, he had rejected the exclusive focus on Christian history, putting his attention on a more universal sense of becoming—"the becoming of everything" as Clayton described it.

As I listened, Clayton's description of Pannenberg called to mind Hegel, who had also understood spirit to be deeply involved with the development of culture and history. Hegel had suggested that God is, in some sense, embedded in the becoming of the entire world. When I asked Clayton about the connection, he recalled that he had suggested the same to his mentor when they first met in 1981. "Herr Pannenberg," he had ventured, "I recognize that your theology is fundamentally Hegelian." Pannenberg immediately bristled at the assumption and replied with a scowl to his new student, "Vat do *you* know of Hegel?"

During Clayton's time in Germany, the existential confusion created by the break with his theological certitude began to settle and he started to find a rich new vision of spirit in the "becoming of everything." His emerging recognition that God is found in the very process of history rather than simply at the pinnacle or the beginning was critical. This insight, combined with his fascination with science and his passion for questions, came together to sow the seeds of a new synthesis—an evolutionary theology that he has spent the better part of his life working out.

Soon Clayton sat down to write his first book—*The Evolution of the Notion of God in Modern Thought*. He wrote in German, and by the time he put pen to paper he had come under the influence of Germany's rich and enduring philosophical tradition. But it was not Hegel's evolutionary vision that captured his heart—it was another German idealist, Friedrich Schelling.

"I worked for eight years on the book," Clayton told me, "and I ended up siding with Schelling. Hegel saw the evolutionary process as controlled by the unfolding of absolute logic . . . absolute spirit is the culmination of the unfolding of that logic. But for Schelling, *no law of logic ties together the moments of evolution*. That's absolutely crucial. There are those thinkers who lay evolution out as if it's preordained in some kind of platonic space. But there are also those of us who think that the creative process is just that—*creative*, spontaneous, unfolding, unpredictable. The future can't be known, because the future doesn't exist. And that's why I'm a Schelling evolutionist and not a Hegel evolutionist."

THE EVOLUTION OF THE EVOLUTION DEBATE

Clayton's point about the open-ended creativity of the evolutionary process struck me, in part, because it brought home and supported one of my own observations regarding the history of evolutionary theory. If you look at the general range of theories about the evolutionary process, beginning in the nineteenth century, and then follow the evolution of the evolution debate, so to speak, over the course of the twentieth century and on into our own time, an important trend stands out. Over time, our understanding of the evolutionary process trends toward theories of development that involve more creativity and agency and that are less deterministic. This is true whether we are talking about a scientific or spiritual interpretation of evolution.

For example, science today is well schooled in the indeterminacies of quantum physics and the unpredictable self-organizing outcomes of complexity theory. There is a growing appreciation for the creative power of nature and how difficult a task it is to peer into the future of the evolutionary process. Yet it has not always been so open-minded. We've come a long way from the days when Laplace thought that entire future of the universe could theoretically be predictable if only we knew the exact properties of every last particle.

Theists as well, until recently, have leaned toward a deterministic view of the universe, or at least the view that all of the unpredictability and novelty was attributed to the will of an omnipotent God. And both Teilhard de Chardin and Sri Aurobindo, while they were far from deterministic and appreciated evolution's creative power, each described metaphysically specific blueprints and outcomes by which they saw the future of the process unfolding—in Teilhard's case, the "Omega Point," and in Aurobindo's case, an ascending series of higher levels of consciousness, from higher mind to supermind and beyond. Contemporary thinkers like Clayton, however, are rejecting such preset evolutionary scripts in favor of open-ended, creative, evolutionary systems where novelty reigns and the future is unknown. The issue looms large in the formation of a new worldview, because the less predetermined the outcomes, the more human choice plays a role in setting the direction of the future. An evolutionary system that embraces greater determinism—whether from the scientific or spiritual side of the spectrum—would inevitably reflect the disturbing consequences and dispiriting conclusions of that bias. If evolution's future is already written, our choices become less consequential. But if evolution's future is unwritten, then all kinds of possibilities remain, and our creative human agency will play no small role in determining the shape of tomorrow. Indeed, as we recognize human choice to be more and more the fulcrum on which the future of evolution depends, the ethical consequences intensify. No higher power or inevitable historical movement can replace the critical importance of human agency.

So we would do well to steer clear of the overconfident certitudes of the past, when the best minds of the day in both science and spirit imagined that evolution played out on a predetermined map. But important questions remain: How far do we go? How free and creative is the evolutionary process? How predictable is our future? Recognizing that nothing is predetermined does not equate to the opposite fallacy—that history is irrelevant and anything at all is possible. Open-ended does not necessarily mean a blank slate, that there is no influence from the past. That's true whether we talk about nature or culture. Indeed, I have seen too many people marry the doctrine of indeterminacy with the power of human choice and arrive at a strange concoction—convinced that we can create the future however we like, that everything is open and can change on a dime. They imagine that we hold an infinite creative power to shape that unformed future with our own agency. This is the position often espoused in popular spiritually oriented interpretations of quantum mechanics, where quantum indeterminacy has become an excuse to imagine a world of unbridled creativity, a universe in which physics will bend to our personal will. I also see this fallacy played out in the popular idea that human culture is preparing for a massive large-scale evolutionary leap, a ubiquitous transformation in which the whole world reaches a higher level of consciousness— all together, all at once. As we've discussed, this kind of thinking is idealism to the point of absurdity, well-intentioned perhaps, but all in service of a misguided understanding of how evolution works in consciousness and within a culture.

So while I agreed with Clayton's nod to the open-ended nature of evolution, I also had questions. After all, as we have seen repeatedly in the course of this book, there are clear patterns, trends, and principles that inform evolutionary unfolding. The evolution of life and consciousness is not purely random or throw-up-your-hands spontaneous. There are clear maps that provide context for the past and clues to the future. There may be more freedom in the process than we ever realized, but there is also constraint. There is, in other words, some form of logic to the process—an open-ended, contingent, creative form, but a logic nonetheless. Schelling might say that evolution is creative, spontaneous, and unpredictable, but then how do we account for the seemingly nonrandom results in both biology and human culture?

When I asked Clayton this question, he qualified his previous statement in an important way—one that echoes the ideas of Peirce, Wilber, Sheldrake, and indeed, Whitehead, as discussed in chapter 11. "There can be grooves within open processes," he explained, acknowledging that open-ended doesn't equate to "anything goes." There can be what Pannenberg used to call a " 'lure from the future that is not determinative.' "

Since I was visiting a center dedicated to Whitehead's philosophy, I reflected further on how he had formulated this issue. In Whitehead's view, each moment of experience, or "occasion" as he referred to it, is being created by the converging moments or occasions that have come before. He writes that the "whole universe is an advancing assemblage of these processes [of experience]." All of the preceding moments of experience cascade into the present and are integrated. Whitehead described this process with the enigmatic phrase "the many become one and are increased by one." That means that the *many* events of the past cascade into the present and converge, creating a *new* moment, and thereby increase their number by one. According to this perspective, the present and future are constantly being created anew as the "whole antecedent world conspires to produce a new occasion."

So without question, the present and future are heavily influenced by the past. But for Whitehead, there is another key factor. At every moment, creativity is possible; the potential for novelty exists. In every cascading occasion of experience, there is the opportunity for something new to exert its influence. "The antecedent environment is not wholly efficacious in determining the initial phase of the occasion which springs from it," writes Whitehead. Believe it or not, that's actually one of his simpler sentences. Basically, it's a way of saying that the future is not entirely determined by the past. We don't live in a deterministic universe. At every moment, potential novelty is present in the struggle to form the future out of the events of history. But notice the dynamic tension between freedom and historical determinism in Whitehead's view—the unformed potential for novelty is in a constant, active relationship with the weighty influence of what's come before. It is a dynamic we can easily see in our own lives as the power of our free will interacts with the influential tendencies of our own established psychological, social, and cultural predilections—the result of which shapes the future destiny of our lives.

We could also say that the potential for novelty appears to increase as evolution moves forward. A plant has more potential for novelty than a molecule has. A chimpanzee has more potential for novelty than a plant has. And a human being has more than a chimpanzee has. There is more novelty in biology than in physics, and more in cultural evolution than biological evolution. Science writer John Horgan stumbled on this truth in a 1995 article on complexity theory, noting that scientists were having trouble applying the mathematical standards of physics to the messier world of biology. "Numerical models work particularly well in astronomy and physics because objects and forces conform to their mathematical definitions so precisely," he writes. "Mathematical theories are less compelling when applied to more complex phenomena, notably anything in the biological realm." I would suggest that part of the problem isn't just more complexity; it's also the increased capacity for novelty and agency that exists in the biological world.* Mathematical models could tell us with a high degree of accuracy where Jupiter will be in ten thousand years, but might have trouble pinpointing where my cat will be in ten minutes.

As evolution proceeds, agency evolves, choice increases, consciousness and freedom expand, and so does our capacity to act creatively, to influence the flow of becoming. That means that I have more freedom from the dictates of my past than my cat does. I am less predictable; my choices express more freedom. My cat certainly has some freedom, some agency, and some capacity to respond as an independent little being with subjective needs, wants, wishes, and feelings. She is far from an automaton—compared to an earthworm, she is a living bastion of freedom. But her agency is limited. Even though

---

* This could also help explain why laws in physics are inherently more determinative than laws in biology, which is a science that provides for much more predictability than general principles in the social sciences, etc. It is also worth noting that Maslow pointed out in his book *Motivation and Personality* that individuals at higher stages of development have more free will than others. This would support the idea presented here and the idea that free will continues to evolve even within human cultural evolution.

I don't know where she will be in ten minutes, I do know that she will be doing one of about six or seven things. She has a limited repertoire compared to her human stewards. Human beings have the potential for tremendous creative agency (not that they always use it) and so our capacity to affect the flow of becoming is that much greater.

One message we can glean from Whitehead's complex thought is this: We always possess the potential to make a leap forward, to liberate our lives from the inertia of the past, to add something new and novel to the march of history, but not to discard it completely. We have a tremendous capacity to mold and shape the future, but not to magically erase what has come before. As we discussed in chapter 3, Evolutionaries must find their way to a deep optimism, grounded in realism. We must steer between a cynical conservatism on one hand, which tells us "there is nothing new under the sun," and a naïve romantic idealism on the other, which tells us that "anything at all is immediately possible." Both are untrue, both deny the actual processes of evolution, and both ultimately impede our capacity to respond effectively to the demands of our world.

EMERGENCE AND OTHER THEOLOGICAL CONUNDRUMS

During his time in Germany, and later at Yale and several other schools, Philip Clayton's interest in the relationship between science and spirit deepened. And the more he understood about the movements of science and philosophy, the more he realized that theology as it was currently constructed was inadequate. Except for a few unconventional thinkers, theology tended to be stuck in the past, unable to engage with contemporary currents of thought and make a coherent case for God that was congruent with the progression of science and philosophy. "Theological reflection on the spirit often contents itself with what is in essence a pre-modern notion of [God]," Clayton writes, "then, when such . . . ways of knowing

God prove to be inadequate or lead to skeptical notions, theologians are tempted to just throw up their hands and declare that Spirit just can't be grasped by the human mind."

This theological gap, Clayton feels, engenders skepticism with religion, as it seems unable to provide a relevant way of making meaning in our contemporary world. Such concerns have inspired him to search for a notion of God that is flexible enough to embrace the extraordinary development of knowledge of the past two centuries—a theological worldview, in other words, that could peer deeply into the natural world as revealed by science, and not flinch.

William Grassie, the director of the science and spirit foundation Metanexus, likes to say that atheists are not particularly openminded about God. In fact, they tend to have a very clear idea of exactly what kind of God they don't believe in. We see this demonstrated in comments made by the New Atheist philosopher Daniel Dennett after seeing a lecture by Clayton: "Clayton astonished me by listing God's attributes: according to his handsomely naturalistic theology, God is not omnipotent, not even supernatural, and . . . in short Clayton is an atheist who won't admit it." I suspect this reveals more about Dennett's narrow conception of God than it does Clayton's supposed atheism. Evolutionary spirituality in all its many forms steers well clear of the supernatural, omnipotent old-man-in-the-sky of tradition. And Clayton is no different. He is bringing forth an understanding of God or Spirit that is quite distinct from the omnipotent deity of his Christian forefathers.

One primary aspect of this new theology comes to us from science: the theory of emergence. For Clayton, and many others, the idea of emergence—a concept that has come to us out of the complexity sciences as well as evolutionary philosophy—has the potential to change the way we think about spirit in a scientifically revealed universe. It suggests that in the process of evolution, fundamentally novel and higher levels of complexity with critical new emergent properties come into existence *that cannot be reduced to the levels below.* The favorite example from chemistry is water. Who could predict water from hydrogen and oxygen? There is nothing contained in those two elements on their own that would lead you to predict such a remarkable offspring from their marriage. It's like two tone-deaf individuals marrying and producing a Mozart: a dramatic, unpredictable, exciting emergence.

"We see in the natural world an open-ended process of increasing complexity, which leads to qualitatively new forms of existence," Clayton writes. "Qualitatively" is the key word here. It suggests that novel modes of being come into existence in the process of evolution that are so fundamentally different from what has come before that their properties cannot be reduced to the other qualities present at a lower level of existence.

Let's again consider my cat, who has been my daily companion in the process of writing this book. When I look at my cat, I see that the matter in her body is quite unique compared to, say, a rock or a molecule or a galaxy or even a tree. But it's still matter—physical stuff. And yet we can see that the qualities of my cat—the incredible powers of movement, her awareness, some form of sentience and consciousness, the ability to recognize her human companions, her playfulness, her ability to maintain some rudimentary form of relationship—are all emergent properties of life. There is just no way to look at trillions and trillions of various atoms in the early constitution of this planet and conclude, "Eventually, we're going to get to cat!" At every major stage of evolution, new, unpredictable emergent properties come into existence.

The idea of emergence is all the more attractive because it passes the "common sense" test. It allows for the idea that things we consider rather important in our everyday lives—for example, free will—are not simply fancy illusions that tantalize us into falsely imagining that life has some vital, mental, or spiritual quality when in fact all is reducible to the interactions of physical particles. No, these emergent stages constitute legitimate novelty, categories of existence that operate with new causal powers and properties. And emergence also suggests a universe that is "upwardly open," as some have referred to it, meaning that there is no reason to think that emergent modes of being will stop at the evolution of human mentality. What new qualities and characteristics does evolution have in store for the next level of emergence? What kind of supramental or transmental categories await us as this unpredictable process unfolds?

At the same time, emergence is one of those ideas that can be exciting one moment and murky the next. It can sometimes be oversold as an answer to the many conundrums that confront us as we look at the trajectory of evolution. I've seen it become a sort of pseudoexplanation of a phenomenon, the new and improved version of "God did it." How did human consciousness evolve? Uh . . . *emergence*! Great, but have we really explained anything?

So while the idea of emergence should not be confused with an explanation of the novelty of nature, it does help identify actual truths about the evolutionary process that are critical to appreciate. It names the wonderful creativity of our cosmic story—radically new capacities and higher levels of being *do* emerge in this marvelous universe. And it brings to light realities that have always been important in both theology and science—such as how and why human beings seem to be unique among nature's inhabitants.

Emergence, then, is one way that we can begin to give legitimacy to the actual qualities present in that awesome sweep of evolutionary unfolding—from matter to life to mind to . . . what? For some theologians, God is the next level in the sequence, the next emergent quality in the natural teleological progression of cosmic evolution in which consciousness, mind, perspective, freedom, agency, and creativity are all deepening in quality and quantity. But clearly, this is not the ancient conception of deity. And it raises more than a few questions. If God exists at the end of the process, what, if anything exists before? If God is being created in the process of evolution, how can he or she be the creator? If God is perfect, as

Saint Thomas Aquinas once argued, how can he or she be involved at all in a process that is always less than perfect?

THE PARADOX OF PERFECTION IN AN UNFINISHED UNIVERSE

One of the great theological and mystical conundrums has always revolved around the perfection of God. In fact, part of the actual spiritual power and beauty of a theistic perspective is the worshiper's experience of a divine presence that is spiritually whole and complete. One might say that much of the intrinsic power of a theistic approach to spirit revolves around that deep relationship between the finite nature of self and the infinite, complete perfection of God "from whom all blessings flow" (as I sang in my childhood church choir). The very essence of the theistic and mystical impulse is expressed in that primordial longing of the individual self for liberation from the limitation, finitude, and partialness of incarnate life. It is a restless longing for completion and perfection, for release from the vicissitudes of opposites in a created world that is always *this* or *that*, suffering and striving without ever arriving. Indeed, great religious figures throughout the ages, including Saint Augustine, have cited this very longing as the best evidence of God's existence. As theologian John Haught expresses so beautifully, "We have a God-shaped hole at the heart of our being." That longing finds its natural resting point in the encounter with its opposite—the "infinite inexhaustible depth" of God's being.

I remember a conversation I once had with a Greek Orthodox elder at a church in Boston. At some point during the several hours we spent discussing the holy life, he looked at me with great intensity and said, "You have to understand that God is *uncreated*." He was expressing a traditional interpretation of deity, in which the divine perfection of the uncreated is not marred by the relative nature of creation. However, a cosmos that is evolving is, by its very nature, the opposite of that uncreated divinity. The world of creation, of time, space, and causation, is neither perfect nor complete but perpetually unfinished. So the challenge for theologians and all those who care about the fate of deity in a scientific age is to explain what the intrinsic and transcendent perfection of God's being has to do with the very imperfect, incomplete world of becoming that we all share. In a sense, this has always been the burden placed on a God-centered view of the world. Only now the challenge is not just to account for a world of transience, impermanence, and change but also for a cosmos that is evolving, that is moving, that is going somewhere. And the challenge of that theological conundrum has only become more acute in a world in which our knowledge of the richness and wondrous beauty of nature seems to increase every day whereas our connection with a transcendent theistic presence seems to simultaneously grow more ephemeral and theoretical.

Adding to this challenge for theology is the orthodox theistic model that suggests that the world is a fallen realm, a mere shadow of the divine—a place that we must suffer through and endure, that tests our moral mettle but is far from the bosom of God. Much of Christian theology was originally influenced by Plato and neo-Platonic thought, which held that the material world was imperfect because it exists in a state of unpredictable flux and change, antithetical to the unchanging order and perfection of God. We should not look to the untrustworthy fickleness of the world as our model for divine contemplation but upward toward the "fixity of the heavens."

These days, instead of inspiring people to contemplate the fixity of the heavens, such a perspective tends to embolden them to leave theism altogether—abandoning the church as a relic, a historical institution that has lost the pulse of spirit in the modern age. And so the theological question remains. What do such notions such as infinity, perfection, and completeness mean in an evolutionary age?

One alternative to the traditional split between God and the world is the outlook of panentheism, a term that Clayton suggests was first used by Schelling. It is the conception that God or divinity is intrinsic to the natural world but is not *limited* to the natural world. God is both immanent and transcendent. Panentheism should not be confused with *pantheism*, the idea that nature *is* God. Many scientists flirt with a pantheistic view of nature, finding a deep sense of reverence and spiritual sustenance in the contemplation of wonders of the natural world. It was Spinoza who said *"Deus sive Natura"*—God is nonseparate from nature. We hear echoes of this view in the musings of the Romantics but also in scientists such as Einstein who said that he believed in Spinoza's God. We also hear it in many of the Evolutionaries who have been mentioned in these pages—Sagan, Swimme, Kauffman, and others.

Panentheism, on the other hand, retains the transcendent quality of God while making room for God's deep mystery to also be revealed in the beauty and majesty of nature. However that still leaves us with a conundrum. How could a world in which God is immanent be imperfect, incomplete, and so full of strife and suffering? Theologians have addressed these questions in many ways over the centuries, but with the dawning of an evolutionary perspective, a much more satisfying understanding of the relationship between an uncreated God and creation reveals itself.

"Something exists as the ground of all things and the ground notion is the most basic metaphysical notion across the world's traditions," suggests Clayton. "Something emerges out of it, which is influenced by that ground, but also brings about a fullness of experience that can't be actualized apart from the evolutionary process." This notion of "ground" has been championed by many philosophers and theologians over the years, including Schelling, but perhaps most notably twentieth-century protestant theologian Paul Tillich, who said that God was not a being but the ground of all being.

But once we leave behind that perfect ground, we enter into the created realm, into the cosmos of time and space, a finite world of limitation, struggle, pain, and suffering. According to evolutionary theology, we are not stepping down into a lesser, shadowy, fallen world far from the spirit of God but instead are moving forward into a new dimension of divinity. We are entering into a vast process of becoming and emergence that is also *not separate from God's being*. God is becoming richer, more complete, more all-embracing through the experience of the universe's becoming. And so the limitations of this manifest world are not so much an indication of its separation from God, as theology once conceived, but an expression of God's own internal desire to become greater, richer, more full and complete. Evil, strife, and suffering are not signs of the absence of divinity so much as they point to the unfinished nature of the created universe. And as subjects in this universe, we each have the capacity to participate in the struggle to actualize God's future being, which transcends and includes our own being, in the emergent, creative processes of evolution. Emergent evolution, we might say, is the trace outline of divinity, successive hints of spirit in the processes of matter, as God's very being develops along the arc of cosmic becoming.

"The full personality of God didn't preexist the world, like the traditions used to teach," Clayton told me. "Rather, I would say that something is not complete in God, the fullness of divine experience is not complete. And so the evolutionary process is launched and God becomes [more] through that process."

Clayton's words reminded me of another unusual champion of the theistic God with whom I'd had the pleasure of spending time some months before—John Haught, a Catholic theologian from Georgetown University. Haught easily speaks the language of faith, God, and belief, but he is also a significant player in the larger project of forging an evolutionary spirituality. Like so many of the voices in this book, he was inspired by Teilhard de Chardin. In fact, of all the Evolutionaries I have met, he may be the closest to representing the great Jesuit's theological vision.

"Teilhard was one of the first scientists in the twentieth century to become aware that the universe is a story," Haught explained to me. "It's not just a place of imperfection but a place of creativity and becoming. This meant that we could no longer look spatially somewhere else to find the perfection that we're looking for. We have to look toward the future. The future became for Teilhard the place where we lift up our eyes and our hearts to have something to aspire to."

Haught takes the *theos* part of his theology very seriously. He expressed his disappointment with more pantheistic evolutionary philosophers who are willing to talk about the immanent divinity in nature but shy away from talk of God and the transcendence that such a word implies. But he also made it clear several times that there is nothing old-fashioned about what he means when he uses this ancient term. For Haught, like Clayton, God is intimately involved in the processes of evolution. "Evolutionary theology suggests that the body of Christ, which in a real sense includes the whole cosmos, is still in the process of being formed," he writes. And mirroring the shift that we saw in the Eastern-inspired traditions of Aurobindo and Cohen, who have similarly redefined the goal of spiritual liberation, he reconfigures the Christian notion of salvation: "Too often, we have thought that Christ's salvific role is that of liberating our souls *from* the universe rather than making us part of the great work of renewing and extending God's creation."

"The world must have a God; but our concept of God must be extended as the dimensions of our world are extended," wrote Teilhard, almost a century ago. He predicted that the religions that would survive would be those that were willing to develop forms of their traditions that organically embrace the reality of an evolutionary worldview. After talking with Haught and Clayton, I think I began to better understand the clarity of Teilhard's foresight. Indeed, just as a God who lives in and through nature might have been the most relevant form of divinity to a hunter-gatherer tribe embedded in the cycles of the natural world thousands of years ago; and just as a transcendent God who offers infinite peace, rest, and redemption, beyond time and the world, might have made perfect sense for the "nasty, brutish, and short" lives of our forefathers; so too does an evolutionary conception of God fit hand-in-glove with the fast-changing, globalizing, rapidly complexifying world of our own time. The consciousness of our age calls out for a divinity that lives not just in the wondrous beauty of nature or the eternal stillness of the present moment but also in the unknown creative potential that exists in the mysterious space of the future. "The future is the primary dwelling place of God," writes Haught. Expanding on this theme in our conversations, he expressed what is perhaps the core idea of an evolutionary theology: "God is not *up above* but rather *up ahead*. In other words, everything that happens in the universe is anticipatory. The world rests on the future. And one could say that God is the one who has future in His very essence."

Some Evolutionaries may always feel that the notion of a God is no longer necessary—discarding deity as an outdated relic of the old static worldview. But for those who feel the age-old pull toward the infinite still tugging at their hearts, who are stirred by the restless longing for ultimacy, these evolutionary theologians offer a deeply satisfying new vision of the divine. They are drawing new and compelling connections between God, the ground of being, evolutionary emergence, consciousness, telos, and the future. Theirs is a God who does not succumb to the spell of solidity, a deity that is evolutionarily inspired, future oriented, and world embracing. And keeping such a God alive in our hearts might be important in saving evolutionary spirituality from its tendency to collapse into pantheism or, in some cases, naturalism. We can touch this form of divinity, not only in the mystical intuition of a transcendent realm of being but in our own efforts to *become*, to give birth to something more good, true, and beautiful in the very processes of the universe's becoming—and ultimately of our own.

CHAPTER SIXTEEN

Pilgrims of the Future

In every great epoch there is some one idea at work which is more powerful than any other, and which shapes the events of the time and determines their ultimate issue.

—Henry Thomas Buckle, introduction to
History of Civilization in England

Recently, I had the opportunity to spend time with the last living person who knew Teilhard de Chardin. Jean Houston is a global teacher, author, storyteller, and Evolutionary all wrapped up into one, and the story she tells of her own initiation into an evolutionary worldview is truly mythic. At the age of fourteen she was living in Manhattan, the grieving daughter of recently divorced parents. One day, while running to school, she accidentally stumbled headlong into an elderly gentleman, who asked her in a thick French accent, "Are you planning to run like that for the rest of your life?" "Yes!" she managed to reply as she ran off down Park Avenue. *"Bon voyage!"* shouted her new acquaintance.

That accidental collision would prove to be a decisive moment in the young woman's life, and set the stage for an unlikely friendship. The next time she ran into this man was the following week, while walking her dog. He recognized her immediately, and they struck up a conversation. It didn't take long for her to realize that this man was no ordinary adult. He had "no self-consciousness," she remembers, and seemed to be "always in a state of wonder and astonishment." Unable to grasp his complex French name, she simply called him "Mr. Tayer." But somehow, her fourteen-year-old mind was perceptive enough to appreciate that she was in the presence of greatness, and the conversations with "Mr. Tayer" were worth remembering and writing down.

He taught her many things on their rambles in Central Park over the next couple of years. She recalls how he filled her young mind with visions of "spirals and nature and art, snail shells and galaxies, the labyrinth on the floor of Chartres Cathedral and the Rose Window and the convolutions of the brain, the whirl of flowers and the circulation of the heart's blood. It was all taken up in a great hymn to the spiraling evolution of spirit and matter."

The last time she saw him was in the spring of 1955, the Sunday before Easter. At one point during the conversation, she worked up the courage to ask him a question about himself. His answer would remain permanently etched in her mind: "I believe that I am a pilgrim of the future," he told her. "Jean, the people of your time, toward the end of this century, will be taking the tiller of the world. Remain always true to yourself, but move ever-upward toward greater consciousness and greater love."

"Those were the last words that he said to me," she recalls. "Then he said, '*Au revoir*.'"

For weeks, she returned to Central Park and waited for him, to no avail. Only years later, when someone gave her a book called *The Phenomenon of Man*, did all the pieces fall into place. There on the back cover was the unmistakable face of Mr. Tayer. Teilhard de Chardin had been her mentor, and he had died Easter Sunday in 1955. She went on to build a remarkable life commensurate with the lessons she learned on those magical walks in Central Park.

Pilgrims of the future. That is a perfect way to describe the Evolutionaries in this book. A pilgrim means a person who comes from afar, traveling on a quest to a sacred place. In this case, that pilgrimage destination is not a physical place but a psychic, cultural, and cosmic possibility—the as-yet-unrealized potential of the future. To be an Evolutionary means to reach out beyond the edges of what has already occurred, to see oneself as journeying into uncreated territory. And I think all the Evolutionaries in this book, whatever their spiritual or religious convictions, would feel at home with that characterization of their life and work.

In the same way that priests once were tasked with interpreting the mythic, metaphysical world, and scientists are culturally designated to help us understand the natural world, I have proposed in these pages that "Evolutionaries" is the best term yet to describe those individuals who feel called to illuminate and interpret the many dimensions of the evolutionary universe that we find ourselves living in.

I hope it has become clear how that activity, that multidimensional grasp of an evolutionary context, always points us toward the future. Remember that I identified the phrase *"we are moving"* as the touchstone proposition of an evolutionary worldview. And what do we do when we find ourselves riding upon something that is in motion? We immediately and instinctively look ahead. We peer out in the direction of movement to see where we are going, what possibilities or challenges lie in our path. When it comes to evolution, however, we are not looking ahead in space, we are looking ahead *in time*. Evolutionaries may dig deep into the past and explore the dynamics of our world from multiple angles, but the very nature of an evolutionary worldview means that the inner compass always comes to rest with its needle pointing toward the future.

There was a time, a couple hundred years ago, when the word "scientist" was not yet a concept formed in the collective mind. The modern activity of science was itself still too new to require such a designation. "Natural philosopher" was the term being used to describe scientists of the day, a phrase that was more rooted in the premodern world and didn't really capture the objective, experimental pursuits of science. Today, of course, the distinction seems obvious; in the early nineteenth century, it was a barely formed intuition.

Whether "Evolutionaries" ultimately captures the mind-space of the culture and survives as a designation, only time will tell. But I hope the reader has been able to appreciate, in the course of this book, just how significant this new lens is through which we are able to examine the nature of consciousness, culture, and cosmos. And I hope that it has become clear how much evolution as an idea has expanded beyond Darwin's framing. The idea has gone viral and escaped the walls of biology, moonlighting in numerous fields, transforming far more than the way we think about fruit flies and fossils. We can certainly still appreciate the time-tested and scientifically verified idea that evolution is happening at the level of the gene. But as these pages have shown, in science, culture, and spirituality, evolution has come to mean much more. As we have progressed through the book, chapter by chapter, we have examined the evolution of cooperation, agency, technology, information, worldviews, consciousness, perspective, creativity, God, and even evolution itself. The sheer pervasiveness of an evolutionary worldview is reflected in that list. Evolution is a promiscuous metaconcept that breaks down intellectual silos and integrates across disciplines. It is truly "a curve that all lines must follow."

Today, we can see how the spell of solidity is being shattered in discipline after discipline. The result has been a long, slow revelation that the ground beneath our feet is moving forward in history. We are in the midst of an epochal shift from a world of stasis to one of constant movement, from a universe of settled being to one of creative becoming, from a cosmos composed of matter in stasis to one made of events in motion. As we add this new sense of temporality to our universe and it becomes more integrated into the patterns of our perception, informing our cultural worldview and restructuring our psychology and neurology, a new sense of the world will reveal itself.

I know that for those who have found the insights in these pages rewarding, sooner or later questions will arise as to how one should go about applying them to a world that is desperately in need of so many of the individual qualities and evolutionary principles described here. I appreciate that impulse. When one begins to internalize the ideas of an evolutionary worldview and authentically question the spell of solidity that centuries of cultural conditioning have cast over our consciousness, the result is powerfully liberating. And the sense of possibility that arises on the other side of such an experience is intoxicating. I would only point out that to actually absorb these ideas is to change one's deepest structures of self, perhaps irrevocably. That does not happen in a day, in a moment of revelation, or in five simple steps. It takes time, consideration, contemplation, deliberation, and introspection. We must be willing to risk our established modes of knowing and seeing the world around us. We must find the courage and authenticity to not settle for superficial bursts of inspiration or temporary flights of insight but to pursue these ideas all the way to the deepest interiors of the self, where new perspectives take root and new worldviews form. If you want to apply these ideas, that's where you have to start.

So many people today have lost faith in the power of deep thinking, in the ability of novel insights and emerging truths to change our hearts and minds, to freshly inspire and radically reorganize our categories of consciousness. And they are often convinced that when it comes to the problems of our global society, fundamentally, we already have the answers. All we lack, they feel, are the practical resources, the institutions, or the collective will and political power to apply them. I understand the frustration, but I would suggest that theirs is the frustration of a static worldview—one that does not allow for the possibility of genuine evolution, either in the world they are seeing or, more important, in the lens through which they are seeing it. I hope this book has begun to challenge such convictions.

I'm convinced that the emergence of an evolutionary worldview has the potential to have a beneficial effect on all levels of society over the next decades and centuries. But just as it took time for the essential ideas of the European Enlightenment to find their way into the lived political and social freedoms we now enjoy so readily, it will take time for the core insights of the evolutionary worldview to develop into the political and social applications that will more directly address the global challenges we currently endure.

In fact, I would suggest that many of the ideas highlighted in this book belong closer to the core of this emerging worldview. Not quite practical, they help define the space in which more pragmatic expressions can be forged over time. There are no doubt many books to be written about the way in which evolutionary ideas will influence psychology, politics, social change, history, economics, law, and many other fields. I look forward with great anticipation to reading those books. Perhaps I will even write one. But the power of those works will still reside in the willingness of the authors to invest the time and personal commitment to care about the truth of these ideas, to break the spell of solidity in themselves so deeply that their own perspectives could never remain the same. Then, and only then, will they be helping to create the very patterns and practices that will help define and build this new worldview.

Like Teilhard, I aspire to be a pilgrim of the future. I have great conviction in the power and potential of this new worldview. I genuinely believe it can dramatically energize our society and provide a pathway into the future that is consonant with the best of human culture. I am not alone in that conviction. Indeed, the most powerful part of writing a book such as this has been the opportunity to spend time in the company of so many others who share a passion for this cultural project. As I mentioned in the very first chapter, there is absolutely nothing "solo" in this endeavor. Building an evolutionary worldview that has authentic cultural resonance in the twenty-first century is a massive undertaking, requiring the help of all those who, inspired by the beauty of an evolutionary vision, have made it a significant part of their life's work. Unlike Teilhard, Huxley, Aurobindo, Whitehead, Gebser, Bergson, Baldwin, and others, today's Evolutionaries are not lone torches, shining brightly in an otherwise pitch-black night. They are part of a larger movement—a fledgling, unstructured, diverse movement, but one with great cultural promise and significance. I hope this book helps galvanize and unify those already shaping this field and inspire a new generation of Evolutionaries to see just how compelling, fulfilling, and culturally relevant the ideas at the heart of this emerging worldview are. In the past century, as Teilhard predicted, we have taken up the tiller of the world. May nature's exuberant creativity guide our hands. In pursuing our passion for the possible we will find the future of evolution.

ACKNOWLEDGMENTS

First, I want to thank my creative partner and publishing advisor Ellen Daly (who also happens to be my wife) for her extraordinary commitment to this project, from proposal to publication, which truly made everything possible. I want to extend my gratitude to my agent, Natasha Kern, whose clearheaded advice and encouragement was invaluable through the ups and downs of the publishing process, and to my masterful editor at Harper Perennial, Peter Hubbard, for believing in this project and recognizing that somewhere in between straight science and popular spirituality are serious messages worthy of attention.

This book would never have come into being without the stimulating, challenging, creative vortex that I have shared with my close friends and colleagues at *EnlightenNext*, helping to nurture the ideas at the heart of an evolutionary worldview. I'm also grateful for the time and space they generously gave me—and the extra work they shouldered—to allow me to focus on this project. I want to express my appreciation for all of the Evolutionaries who so graciously shared their inspiring thoughts, stories, and visions with me in the research for this book. It is also important to acknowledge that the title of this book was a product of the remarkable foresight and branding wizardry of Kevin Clark, who (along with a number of others) independently coined the term "Evolutionaries" and introduced me to it. Another note of gratitude goes out to Patrick Bryson, whose creative input and advice was critical in arriving at a satisfying cover design. There were also many individuals who graciously donated their time and energy to review the manuscript and provide feedback—including Elizabeth Debold, Steve McIntosh, Connie Barlow, Tom Huston, Ross Robertson, and Michael Dowd. The generosity of Melissa Hoffman was important to this book, and I'm appreciative of the secluded writing space she provided at critical junctures along the way.

Finally, I want to thank two formative influences on my life and work. First, my mother, Mona Phipps, and my late father, Kent Phipps, who always encouraged me to think seriously about the most important questions of life and graciously gave me the space to pursue those questions, even down unconventional paths. And second, my spiritual teacher and mentor, Andrew Cohen, for his constant support and enthusiasm, and for sparking an evolutionary fire in this heart that has only burned brighter over the many years of our work and friendship.

NOTES

INTRODUCTION

x **If the pollsters are to be trusted:** *Newsweek* poll conducted by Princeton Survey Research Associates International, *Newsweek*, March 28–29, 2007.

xii **"a slow hunch":** Johnson, Steven, *Where Good Ideas Come From* (New York: Riverhead, 2010), 78.

PROLOGUE

4 **"Evolutionary man can no longer take refuge":** Huxley, Julian, "The Evolutionary Vision" in Tax, Sol, and Charles Callender, eds., *Evolution After Darwin* (Chicago: University of Chicago Press, 1960), 252–253, 260.

CHAPTER I

8 **When the historian Will Durant was asked:** Durant, Will, *The Greatest Minds and Ideas of All Time*, ed. John Little (New York: Simon & Schuster, 2002), 1.

11 **"Nothing makes sense in biology":** Dobzhansky, Theodore, *American Biology Teacher* Vol. 35, No. 3 (March 1973), 125–129.

11 **"Is evolution a theory":** Teilhard de Chardin, Pierre, *The Phenomenon of Man* (New York: Harper Perennial, 2008), 219.

12 **"A philosophy of this kind":** Bergson, Henri, *Creative Evolution* (Mineola, N.Y.: Dover Publications, 1998), xiv.

13 **"There is considerable doubt":** Kauffman, Stuart, *Reinventing the Sacred* (New York: Basic Books, 2008), 76.

17 **"While we postmoderns say we detest":** Brooks, David, "The Age of Darwin," *New York Times*, April 16, 2007.

20   **"You'll have an explosion"**: in Huston, Tom, "Looking Back to the Beginning," *What Is Enlightenment?* Issue 26, August–October 2004, 27.

CHAPTER 2

22   **"What our awareness delivers"**: Wilber, Ken, "God's Playing a New Game," *What Is Enlightenment?* Issue 33, June–August 2006, 69.

22   **"A world view is a system"**: Aerts, D. et al., *World Views: From Fragmentation to Integration* (Brussels: VUB Press, 1994), accessed October 2011, www.vub.ac.be/CLEA/pub/books/worldviews.

23   **"like the foundations of a house"**: Wright, N. T., *The New Testament and the People of God* (Minneapolis: Fortress Press, 1992), 125.

24   **"At the center of every worldview"**: Halverson, William H., *A Concise Introduction to Philosophy* (New York: Random House, 1976), 384.

25   **"The conflict dates from the day"**: Teilhard de Chardin, Pierre, *The Future of Man* (New York: Image Books, 2004), 1.

28   **"Life in general is mobility"**: Bergson, Henri, *Creative Evolution* (Mineola, N.Y.: Dover Publications, 1998), 128.

29   **"Fallacy of misplaced concreteness"**: Whitehead, Alfred North, *Science and the Modern World* (New York: Free Press, 1997), 51.

29   **"Permanence has fled"**: Eisendrath, Craig, *At War With Time* (New York: Allworth Press, 2003), 243.

CHAPTER 3

33   **"Most educated people"**: Eisendrath, Craig, *At War With Time* (New York: Allworth Press, 2003), 106.

34   **"biobabble"**: Krugman, Paul, "The Power of Biobabble," *Slate* magazine, October 24, 1997.

34   **"to speak for the culture as a whole"**: Eisendrath, *At War With Time*, 107.

35   **"Great disembedding"**: Taylor, Charles, *A Secular Age* (Cambridge, MA: Harvard University Press, 2007), 146–158.

37   **"Nothing in particular"**: in Solomon, Robert C., *In the Spirit of Hegel* (New York: Oxford University Press, 1983), 338.

37   **"There is also a growing need"**: in Christian, David, *Big History* (Berkeley: University of California Press, 2005), 3–4.

38   **"Reductionism alone is not adequate"**: Kauffman, Stuart, *Reinventing the Sacred* (New York: Basic Books, 2008), 3.

39   **"The overlapping domains"**: Gardner, James N., *Biocosm* (Makawao, Hawaii: Inner Ocean Publishing, 2003), 226.

40　**"Our concern is with integrality"**: Gebser, Jean, *The Ever-Present Origin* (Athens, Ohio: Ohio University Press, 1986), 3.

40　**"Simple sudden synthesis"**: Joyce, James, *Stephen Hero* (New York: New Directions, 1963), 212.

42　**"Just as we separate in space"**: Bergson, Henri, *Creative Evolution* (Mineola, N.Y.: Dover Publications, 1998), 163.

CHAPTER 4

50　**Margulis's work on this new theory**: Sagan, Lynn, "On the Origin of Mitosing Cells," *Journal of Theoretical Biology* 14 (1967), 3:255–274.

50　**"Animals are very tardy"**: Margulis, Lynn, "Gaia Is a Tough Bitch," in Brockman, John, ed., *The Third Culture: Beyond the Scientific Revolution* (New York: Simon & Schuster, 1995), 130.

51　**"My major thrust"**: Ibid., 136.

51　**"Three billion years of non-events"**: Wright, Karen, "When Life was Odd," *Discover*, March 1997, 53.

52　**"a minor twentieth-century religious sect"**: in Mann, C., "Lynn Margulis: Science's Unruly Earth Mother," *Science* 252, 380.

52　**"The tiny archaebacteria"**: Sahtouris, Elisabet, "The Wisdom of Living Systems," *What Is Enlightenment?* Issue 23, Spring/Summer 2003, 20.

53　**"Before our new wave of knowledge"**: Sahtouris, Elisabet, "The Evolving Story of our Evolving Earth," paper presented at the Foundation for the Future workshop "How Evolution Works" (Seattle, Wash., November 4 and 5, 1999), accessed October 2011, http://www.ratical.org/LifeWeb/Articles/H3Kevolv.

54　**"The aesthetic beauty of these"**: Ben-Jacob, Eshel, "Bacteria Harnessing Complexity," *Biofilms* Vol. 1, Issue 4, Oct. 2004 (Cambridge, UK: Cambridge University Press), 241.

57　**"toxic pollutant holocaust"**: Bloom, Howard, *The Global Brain* (New York: Wiley, 2000), 22.

57　**"How did the eukaryotic cell"**: Margulis, "Gaia Is a Tough Bitch," 137.

62　**"Since the origin of evolutionary biology"**: Roughgarden, Joan, *The Genial Gene: Deconstructing Darwinian Selfishness* (Berkeley, Calif.: University of California Press, 2009), 235.

62　**"We are survival mechanisms"**: in Broom, Donald, M., *The Evolution of Morality and Religion* (Cambridge, UK: Cambridge University Press, 2003), 197.

63　**"Social selection"**: Roughgarden, *The Genial Gene*, 61.

64 **"You humans, when are you going to learn":** *Men in Black*, directed by Barry Sonnenfeld, Amblin Entertainment, 1997.

65 **"So sensitized have I been by":** Thompson, William Irwin, *Coming into Being* (New York: Palgrave MacMillan, 1998), 18.

67 **"Evolution is no linear":** Margulis, Lynn, and Dorian Sagan, *What Is Life?* (Berkeley, Calif.: University of California Press, 2000), 93.

CHAPTER 5

74 **"That biology can be co-opted":** Morris, Simon Conway, *Life's Solution* (Cambridge, UK: Cambridge University Press, 2004), 323.

74 **"The prestige of evolutionary research":** Mayr, Ernst, letter to G. G. Ferris, March 28, 1948, in Ruse, Michael, and Joseph Travis, *Evolution: The First Four Billion Years* (Cambridge, Mass.: Belknap Press, 2009), 35.

76 **"The twentieth century was a great burial ground":** Salvadori, Massimo, *Progress: Can We Do Without It?* (London: Zed Books, 2008), 99.

76 **"He quickly learned":** Wilson, David Sloan, *Evolution for Everyone* (New York: Delacorte Press, 2007), 191.

77 **"a noxious, culturally embedded":** Gould, Stephen J., in Nitecki, M. H., ed., *Evolutionary Progress* (Chicago: University of Chicago Press, 1988), 319.

78 **"Rerun the tape of life":** Morris, *Life's Solution*, 282.

82 **"Cooperation emerges only when evolution":** Stewart, John, "The Evolutionary Manifesto: Our Role in the Future Evolution of Life" (June 6, 2008): 8, accessed September 2010, http://www.evolutionary manifesto.com/man.pdf.

85 **"intentional evolutionaries":** Ibid.

86 **"the near eradication":** Ibid.

87 **600,000 autonomous polities**: Wright, Robert, *Nonzero: The Logic of Human Destiny* (New York, N.Y.: Vintage, 2001), 209.

89 **It is rare for two countries with Golden Arches:** Friedman, Thomas, *The Lexus and the Olive Tree* (New York: Farrar, Straus and Giroux, 1999), 249.

89 **"infrastructure for a planetary first":** Wright, Robert, *Nonzero: The Logic of Human Destiny* (New York: Vintage, 2001), 332.

89 **"Historically, the amity, or goodwill":** Wright, Robert, "The Globalization of Morality," *What Is Enlightenment?* Issue 26, August–October 2004, 36.

90 **"at least suggestive of purpose":** Wright, Robert, "Suggestions of a Larger Purpose," *What Is Enlightenment?* Issue 21, Spring/Summer 2002, 167.

90   **"If directionality is built into life":** Wright, *Nonzero*, 4.

90   **"My parents were creationists":** "Evolutionary Theology," interview of Robert Wright by Deborah Solomon in *The New York Times Magazine*, May 29, 2009, MM22 (New York edition).

91   **"Creationism for Liberals":** Coyne, Jerry A., review of *The Evolution of God*, by Robert Wright, *The New Republic*, August 12, 2009.

92   **"Evolution meanders more than it progresses":** Murphy, Michael, and George Leonard, *The Life We Are Given* (New York: Tarcher, 2005), 170.

97   **"Prehistoric warfare was common and deadly":** LeBlanc, Steven, *Constant Battles: Why We Fight* (New York: St. Martin's Griffin, 2004), 8.

97   **"The world is too much with us":** Wordsworth, William, *The Major Works Including The Prelude*, ed. Stephen Gill (Oxford, UK: Oxford University Press, 2008), 270.

98   **Scholarly evidence . . . matriarchy:** Eisler, Riane, *The Chalice and the Blade* (San Francisco: Harper One, 1988), 59–77.

99   **"dispelling the notion that war is natural":** Eisler, Riane, "The Chalice or the Blade: Choices for Our Future," *New Renaissance Magazine* 7, no. 1 (1997).

99   **"Because we can look back and see the pattern":** "A Song that Goes On Singing," interview of Beatrice Bruteau by Amy Edelstein and Ellen Daly, *What Is Enlightenment?* Issue 21, Spring/Summer 2002, 55.

CHAPTER 6

101   **"Why are there beings at all":** Heidegger, Martin, *Introduction to Metaphysics* (New Haven: Yale University Press, 2000), 1.

102   **"as Gods":** Brand, Stewart, *Whole Earth Discipline: An Ecopragmatist Manifesto* (New York: Viking, 2009), 1.

109   **"How many kinds of atoms":** Bloom, Howard, *The God Problem: How a Godless Cosmos Creates* (Amherst, N.Y.: Prometheus Books, 2012).

113   **"The art of evolution":** Kelly, Kevin, *Out of Control* (New York: Basic Books, 1995), 401.

115   **"precisely defining complexity":** Wright, Robert, *Nonzero: The Logic of Human Destiny* (New York: Vintage, 2001), 344–45.

115   **"How did the beautifully intricate":** Gardner, James N., *Biocosm* (Makawao, Hawaii: Inner Ocean Publishing, 2003), 50.

116   **"powerful idea that order in biology":** Kauffman, Stuart, *Reinventing the Sacred* (New York: Basic Books, 2008), 101.

116   **"ceaseless creativity":** Ibid., 2.

116   **"My claim is not simply that we lack":** Ibid., 5.

117   **"one of the more important":** Wolfram, Stephen, *A New Kind of Science* (Champaign, Ill.: Wolfram Media, 2002), 2.

117   **"locked into the very logic":** Gardner, *Biocosm*, 202.

119   **"If your theory is found to be against":** Eddington, Sir Arthur, *The Nature of the Physical World* (Whitefish, Mont.: Kessinger Publishing, 2005), 74.

120   **"You have to hand it to the creationists":** in Wallis, Claudia, "Evolution Wars," *Time*, Sunday, August 7, 2005.

121   **"A system performing a given basic function":** Dembski, William, *No Free Lunch: Why Specified Complexity Cannot Be Purchased Without Intelligence* (New York: Rowman & Littlefield, 2001), 285.

122   **"What I object to is the narrowness":** Haught, John, "God After Darwin: Haught Response to Behe" on Metanexus.net, December 10, 1999.

123   **"When an embryo begins to develop":** "A New Dawn for Cosmology," interview of James Gardner by Carter Phipps, *What Is Enlightenment?* Issue 33, June–August 2006, 48.

CHAPTER 7

126   **The term "transhumanism" was coined:** Huxley, Julian, "Transhumanism," in *New Bottles for New Wine* (London: Chatto & Windus, 1957), 13–17.

127   **"They damaged his nervous system":** Gibson, William, *Neuromancer* (New York: Ace, 1984), 6.

128   **"marriage of the born and the made":** Kelly, Kevin, *Out of Control* (New York: Basic Books, 1995), 2.

130   **"an exponential runway":** Vinge, Vernor, "The Coming Technological Singularity" (1993), accessed September 2011, http://www.rohan.sdsu.edu/faculty/vinge/misc/singularity.html.

131   **"The prospect of building godlike creatures":** de Garis, Hugo, *The Artilect War* (Palm Springs, Calif.: Etc Publications, 2005), 1.

131   **Some have traced it . . . John von Neuman**: In 1958, Stanislaw Ulam mentions the term in reference to a conversation with John von Neumann: Ulam, S., "Tribute to John von Neumann," *Bulletin of the American Mathematical Society*, 64, no. 3 (May 1958), 1–49.

132   **"on the edge of change comparable";** Vinge, "The Coming Technological Singularity."

132   **"transform every institution and aspect":** Kurzweil, Ray, *The Singularity Is Near* (New York: Penguin, 2006), 7.

132   **"The singularity will represent the culmination":** Ibid., 9.

133 **"The truth-is-stranger-than-fiction factor":** *No Maps for These Territories*, directed by Mark Neale, Mark Neale Productions, 2000.

136 **"Most technology forecasts and forecasters":** Kurzweil, *The Singularity Is Near*, 14.

140 **"There is only one time in the history":** Kelly, Kevin, "We Are the Web," *Wired*, August 2005.

141 **"No one can deny that a network":** Teilhard de Chardin, Pierre, *The Future of Man* (New York: Image Books, 2004), 165.

141 **"consensual hallucination . . . colorless void":** Gibson, *Neuromancer*, 5.

143 **"The implications of the Law of Accelerating Returns":** Kurzweil, Ray, *The Age of Spiritual Machines* (New York: Penguin, 2000), 260.

144 **"Conceptually, at least, biology":** Arthur, W. Brian, *The Nature of Technology: What It Is and How It Evolves* (New York: Free Press, 2009), 208.

146 **"double aspects . . . conception of the world":** Chalmers, David J., "Facing Up to the Problem of Consciousness," *Journal of Consciousness Studies* 2 (1995), 3:200–219.

146 **"The quiet, unobtrusive way":** Haught, John, *Is Nature Enough?* (Cambridge, UK: Cambridge University Press, 2006), 68.

Chapter 8

156 **"While I think natural selection":** "Suggestions of a Larger Purpose," interview of Robert Wright by Elizabeth Debold, in *What Is Enlightenment?* Issue 21, Spring/Summer 2002, 106.

157 **"Motive principle of evolution":** Bergson, Henri, *Creative Evolution* (Mineola, N.Y.: Dover Publications, 1998), 182.

160 **"living tissue of shared experience":** Houston, Jean, from an *EnlightenNext* webcast dialogue with Andrew Cohen (September 2011), accessed October 2011, http://www.evolutionaryenlightenment.com/webcast.

165 **That distinction goes to Russian scientist Vladimir Vernadsky:** Vernadsky, Vladimir, "The Biosphere and the Noosphere," *American Scientist*, January 1945, 1–12.

165 **"The Physical and the Psychic":** Teilhard de Chardin, Pierre, *The Future of Man* (New York: Image Books, 2004), 209.

165 **Law of complexity and consciousness:** Teilhard de Chardin, Pierre, *The Phenomenon of Man* (New York: Harper Perennial, 2008), 61.

166 **"Existence mysteriously becomes experience":** Godwin, Robert, *One Cosmos Under God* (St. Paul, Minn.: Paragon House, 2004), 19.

166 **"Why, all of a sudden":** Ibid., 101.

167   **"Everywhere the active phyletic lines":** Teilhard de Chardin, *The Phenomenon of Man*, 160.

167   **"a pioneering hominid":** Swimme, Brian, *The Hidden Heart of the Cosmos* (Maryknoll, N.Y.: Orbis Books, 1999), 56.

167   **"a luminous fissure":** Godwin, *One Cosmos Under God*, 56.

167   **"Abstraction, logic, reasoned choice":** Teilhard de Chardin, *The Phenomenon of Man*, 160.

168   **"psychic system of a collective":** Jung, Carl G., *The Archetypes and the Collective Unconscious* (Princeton, N.J.: Princeton University Press, 1968), 43.

169   **"How on earth do you get in my mind":** Wilber, Ken, *Integral Spirituality* (Boston, Mass.: Shambhala, 2007), 151.

169   **"Relationships exist in the internal space":** McIntosh, Steve, *Integral Consciousness and the Future of Evolution* (St. Paul, Minn.: Paragon House, 2007), 19.

170   **"the meanings and rules":** "Wanted: Chief Culture Officer," an interview with Grant McCracken by *Entrepreneur* magazine, June 2010.

C H A P T E R 9

182   **"Consciousness is generally seen":** Sleutels, Jan, "Recent Changes in the Structure of Consciousness?" talk given at "Towards a Science of Consciousness," Tucson, Ariz., April 2008, paper subsequently published in Hameroff, Stuart, ed., *Toward a Science of Consciousness 2008: Consciousness Research Abstracts* (2008), 172–173.

183   **"The characters of the Iliad":** Jaynes, Julian, *The Origin of Consciousness in the Breakdown of the Bicameral Mind* (New York: Mariner Books, 2000), 72.

185   **"One developmental-logical stage":** Owen, David S., *Between Reason and History* (Albany, N.Y.: SUNY Press, 2002), 102.

188   **"dynamic equilibrium":** see Kegan, Robert, "Epistemology, Fourth Order Consciousness, and the Subject-Object Relationship," *What is Enlightenment?* Issue 22, Fall/Winter 2002, 149.

188   **"I wish to write a history not of wars":** in Durant, Will, *The Story of Philosophy* (New York: Simon & Schuster, 2005), 169.

189   **"the Truth is not only the result":** in Solomon, Robert C., *In the Spirit of Hegel* (New York: Oxford University Press, 1983), 245.

189   **"Hegel was the first to recognize":** McIntosh, Steve, *Integral Consciousness and the Future of Evolution* (St. Paul, Minn.: Paragon House, 2007), 161.

189   **"both a valid truth unto itself":** quoted in Solomon, *In the Spirit of Hegel*, 245.

190 **"The bud disappears as the blossom":** in Ibid., 241–2.

191 **Popper went so far as to accuse:** Popper, Karl, *The Open Society and Its Enemies* (Philadelphia, Pa.: Psychology Press, 2003), 66. Written during World War II, Popper's influential book criticizes theories in which history unfolds according to universal laws, and indicts as totalitarian Plato, Hegel, and Marx.

191 **"With Hegel's decline there passed":** Tarnas, Richard, *The Passion of the Western Mind* (New York: Ballantine Books, 1993), 383.

192 **"the branch of social science":** McIntosh, Steve, *Evolution's Purpose: An Integral Interpretation of the Scientific Story of Our Origins* (New York, N.Y.: Select Books, 2012).

193 **"evolution was an absolute conceptual anchor":** Plotkin, Henry, *Evolutionary Thought in Psychology: A Brief History* (Hoboken, N.J.: Blackwell Publishing, 2004), 70.

194 **"The laws of thought":** in Ibid., 72.

194 **"general biology is today mainly":** Baldwin, James Mark, *Development and Evolution* (New York: Macmillan, 1902), vii.

195 **"The great glory within my own field":** "Epistemology, Fourth Order Consciousness, and the Subject-Object Relationship," interview of Robert Kegan by Elizabeth Debold in *What Is Enlightenment?* Issue 22, Fall/Winter 2002, 149.

196 **"Even though individual development":** McIntosh, *Evolution's Purpose: An Integral Interpretation of the Scientific Story of Our Origins* (New York, N.Y.: Select Books, 2012).

198 **"Unlike paleontology, where the outlines":** Lachman, Gary, *A Secret History of Consciousness* (Great Barrington, Mass.: Lindisfarne Books, 2003), 97.

198 **"The imprint of human imagination":** Ibid., 103.

198 **"felt the beat of consciousness":** Richards, Robert J., *Darwin and the Emergence of Evolutionary Theories of Mind and Behavior* (Chicago: University of Chicago Press, 1989), 480.

198 **"brilliantly intuitive intellectual mystic":** Thompson, William Irwin, *Coming into Being* (New York: Palgrave MacMillan, 1998), 12.

198 **He called this consciousness "integral" . . . "authentic spell-casting":** Gebser, Jean, *The Ever-Present Origin* (Athens: Ohio University Press, 1986), 36–102.

204 **"Sing, O goddess, the anger of Achilles, son of Peleus":** Homer, *The Iliad*, trans. Anthony Verity (New York: Oxford University Press, 2011), 3.

208 **"transfiguration and irradiation":** in Lachman, Gary, *A Secret History*

of Consciousness (Great Barrington, Mass.: Lindisfarne Books, 2003), 229–230.

CHAPTER 10

214 **"People are trapped in history"**: Baldwin, James Mark, *Notes of a Native Son* (Boston, Mass.: Beacon Press, 1984), 163.

215 **"The error which most people make"**: Graves, Clare, "Human Nature Prepares for Momentous Leap," edited with embedded comments by Edward Cornish, World Future Society, in *The Futurist*, April 1974, 72–87.

217 **"Briefly, what I am proposing"**: Ibid.

218 **Spiral Dynamics value systems**: Beck, Don, "The Never-Ending Upward Quest," *What Is Enlightenment?* Issue 22, Fall/Winter 2002, 105–126.

218 **"the first sense of the metaphysical"**: Beck, "The Never-Ending Upward Quest," 113.

219 **"ritualistically and superstitiously"**: Graves, Clare, "Dr. Graves's 1982 Seminar Handout: What the Research of Clare W. Graves Says a Model of Healthy Mature Psychosocial Behavior Should Represent," prepared by Chris Cowan, accessed September 2011, http://www.clarewgraves.com/articles_content/1982_handout/1982_1.html.

219 **"raw, egocentric self"**: Beck, "The Never-Ending Upward Quest," 114.

220 **"Spiral Dynamics is based on the assumption"**: Beck, "The Never-Ending Upward Quest," 110.

220 **"not rigid levels, but flowing waves"**: Wilber, Ken, *A Theory of Everything* (Boston, Mass.: Shambhala Publications, 2001), 7.

221 **"What I am saying is that when one form"**: in Beck, Don Edward, and Christopher Cowan, *Spiral Dynamics* (Hoboken, N.J.: Wiley-Blackwell, 1996), 294.

222 **"We do not need a time machine"**: Godwin, Robert, *One Cosmos Under God* (St. Paul, Minn.: Paragon House, 2004), 166.

CHAPTER 11

236 **"One thing was very clear to me"**: Wilber, Ken, *Sex, Ecology, Spirituality*, introduction to the revised edition (Boston, Mass.: Shambhala, 2001), xii-xiii.

238 **"There were linguistic hierarchies"**: Ibid., xiii.

239 **At one point, I had over two hundred"**: Ibid.

244 **"The real intent of my writing"**: Wilber, Ken, introduction to vol-

ume 8 of the *Collected Works* (Boston, Mass.: Shambhala Publications, 2000), 49.

245 **"this does not mean that development"**: Wilber, Ken, *A Theory of Everything* (Boston, Mass.: Shambhala Publications, 2001), 22.

246 **"In every work of genius"**: Emerson, Ralph Waldo, "Self Reliance," *Emerson's Essays* (New York: Harper Perennial, 1981), 32.

247 **"replaces *perceptions* with *perspectives*"**: Wilber, Ken, *Integral Spirituality* (Boston, Mass.: Shambhala, 2007), 42.

249 **a fallacy endemic to introspective traditions**: Wilber, *Integral Spirituality*, 272–283.

252 **"May not the laws of the universe"**: Davis, Ellery W., "Charles Peirce at Johns Hopkins," *The Mid-West Quarterly*, September 1914, 53.

253 **"Although cosmology is now evolutionary"**: Sheldrake, Rupert, *Morphic Resonance* (Rochester, Vt.: Inner Traditions/Bear & Co., 2009), xiii.

254 **"According to this hypothesis"**: Ibid., 3–4.

254 **"best candidate for burning"**: Sir John Maddox, "A Book for Burning?" *Nature* 293 (5830), 245–46.

255 **"Kosmic habits"**: Wilber, Ken, "Excerpt A: An Integral Age at the Leading Edge, Part II: Kosmic Habits as Probability Waves," in *Excerpts from Volume 2 of the Kosmos Trilogy*, accessed October 2011, http://wilber.shambhala.com.

255 **"In historical unfolding"**: Wilber, Ken, "Excerpt A: An Integral Age at the Leading Edge, Part I. Kosmic Karma: Why is the Present a Little Bit Like the Past?," in Ibid.

256 **"Most of the early stages"**: Wilber, Ken, "Higher Integration," dialogue with Andrew Cohen in *What Is Enlightenment?* Issue 29, June–August 2005, 58–59.

256 **"This does not mean"**: Wilber, Ken, "Excerpt D: The Look of a Feeling, Part IV: Conclusions of Adequate Structuralism," in *Excerpts from Volume 2 of the Kosmos Trilogy* accessed October 2011, http://wilber.shambhala.com.

257 **"frothy, chaotic, wildly creative"**: Wilber, "Excerpt A: An Integral Age at the Leading Edge, Part II."

258 **"Paul Tillich said that"**: Wilber, "Higher Integration," 59.

CHAPTER 12

264 **"Everyone appeared in high spirits"**: Darwin, Charles, *The Voyage of the Beagle* (New York: Penguin Classics, 1989), 331–332.

264 **"emotionally compelling dramas":** Wade, Nicholas, *The Faith Instinct* (New York: Penguin, 2010), 78.

265 **"Religion is not inserted":** Kardong, Kenneth V., *Beyond God* (Amherst, N.Y.: Humanity Books, 2010), 173.

267 **Fowler's Stages of Faith:** Fowler, James, *Stages of Faith: The Psychology of Human Development and the Quest for Meaning* (San Francisco: Harper One, 1995).

273 **"As I looked at this scene":** Teilhard de Chardin, Pierre, Letter to Marguerite Teilhard, August 23, 1916, in *The Making of a Mind: Letters from a Soldier-Priest* (New York, N.Y.: Harper & Row, 1965), 119–120.

274 **"You seem to feel . . ."** Teilhard de Chardin, Pierre, "Nostalgia for the Front," in *The Heart of Matter* (New York: Mariner Books, 2002), 155.

274 **"Blessed be you, harsh matter":** Teilhard de Chardin, Pierre, "Hymn to Matter," in Ibid., 75–77.

276 **"two negations" . . . "one of the most powerful and convincing experiences":** Aurobindo, Sri, *The Life Divine* (Pondicherry, India: Sri Aurobindo Ashram Trust, 1970), 6–32.

282 **"the most striking feature":** Delbanco, Andrew, *The Real American Dream* (Cambridge, Mass.: Harvard University Press, 1999), 92.

CHAPTER 13

285 **"The last thing I saw with complete clarity":** Torey, Zoltan, *Out of Darkness* (New York: Picador, 2003), 11.

289 **"a vision and a direction":** Hubbard, Barbara Marx, "What Is Conscious Evolution?" accessed October 2011, http://www.barbaramarx hubbard.com/site/node/8.

295 **"Understanding the unwanted drives":** Dowd, Michael, *Thank God For Evolution* (New York: Penguin, 2009), 162.

295 **"New truths no longer spring":** Ibid., 65.

301 **"I lapsed into a day-dream":** Hubbard, Barbara Marx, *The Hunger of Eve* (Greenbank, Wash.: Great Path Publishing, 1989), 66–70.

303 **"a new sense of identity":** Hubbard, Barbara Marx, *Conscious Evolution* (Novato, Calif.: New World Library, 1998), 58.

303 **"natural but dangerous stage":** Ibid., 67.

303 **"reactive response to proactive choice":** Ibid., 68.

303 **"guide their capacities":** Ibid., 69.

305 **"Total Perspective Vortex":** Adams, Douglas, *The Hitchhiker's Guide to the Galaxy* (New York: Del Rey, 2002), 194–198.

305 **"It's really simple":** Swimme, Brian, "Comprehensive Compassion," *What Is Enlightenment?* Issue 19, Spring/Summer 2001, 40.

307 **"Not only are we in the universe"**: Tyson, Neil deGrasse, "Beyond Belief: Science, Reason, Religion and Survival," from a talk given at the Salk Institute for Biological Studies, November 7, 2006, accessed September 2011, http://thesciencenetwork.org/programs/beyond-belief-science-religion-reason-and-survival/session-10-2.

307 **"He listened carefully as I tried"**: Swimme, Brian, foreword to Teilhard de Chardin, Pierre, *The Human Phenomenon* (Eastbourne, UK: Sussex Academic Press, 1999), xiii–xiv.

308 **"We find ourselves ethically destitute"**: Berry, Thomas, *The Great Work: Our Way Into the Future* (New York: Broadway, 2000), 104.

309 **"The fundamental categories of my mind"**: Swimme, foreword to *The Human Phenomenon*, xv.

311 **"Take the discovery of cosmic evolution"**: Swimme, Brian, in an EnlightenNext webcast interview, May 2010.

312 **"The Earth wants to come"**: Swimme, Brian, "The New Story," accessed October 2011, http://www.youtube.com/watch?v=TRykk_0ovI0.

312 **"We happen to be in that moment"**: Swimme, "Comprehensive Compassion," 38.

313 **"It's amazing to realize"**: Ibid., 39.

313 **"Some wellspring of creation"**: Kauffman, Stuart, *Investigations* (New York: Oxford University Press, 2000), 49.

CHAPTER 14

323 **"Existence or Consciousness"**: Maharshi, Ramana, ed. David Goodman, *Be As You Are: The Teachings of Sri Ramana Maharshi* (New York: Penguin, 1989), 15–16.

325 **"I hazard the prophecy"**: Whitehead, Alfred North, *Adventures of Ideas* (New York: Simon & Schuster, 1967), 33.

326 **"became silent as a windless air"**: Aurobindo, Sri, *Letters on Yoga, Volume 3* (Silver Lake, Wis.: Lotus Press, 1988), 172.

327 **"precisely the experience"**: Heehs, Peter, *The Lives of Sri Aurobindo* (New York: Columbia University Press, 2008), 144.

327 **"It threw me suddenly"**: Aurobindo, Sri, *On Himself* (Twin Lakes, Wis.: Lotus Light Publications, 1972), 101.

328 **"feel out for the thought of the future"**: in Heehs, *The Lives of Sri Aurobindo*, 262.

328 **"The animal is a living laboratory"**: Aurobindo, Sri, *The Life Divine* (Pondicherry, India: Sri Aurobindo Ashram Trust, 1970), 3–4.

329 **"the better organized the form"**: in Bruteau, Beatrice, *Evolution To-*

ward Divinity: Teilhard de Chardin and the Hindu Traditions (Wheaton, Ill.: The Theosophical Publishing House, 1974), 157.

329 **"primary divine necessity":** Aurobindo, *The Life Divine*, 47.

329 **"By attaining to the unborn":** Ibid., 48.

330 **"know the cosmic forces as part of himself":** Ibid., 1012.

330 **"This calls for the appearance":** Ibid., 1069.

333 **"I'd never really come across":** Cohen, Andrew, "The Evolution of Enlightenment," a dialogue with Ken Wilber, in *What Is Enlightenment?* Issue 21, Spring/Summer 2002, 42.

334 **"I believe that those of us":** Cohen, Andrew, *Evolutionary Enlightenment* (New York: Select Books, 2011), 4.

335 **"Free, in this sense, means available":** Ibid., 128.

CHAPTER 15

341 **"fog of metaphysics":** Russell, Bertrand, *History of Western Philosophy* (Philadelphia, Pa.: Psychology Press, 2004), 744.

342 **"Bertie thinks I'm muddleheaded":** in Weiss, Paul, "Recollections of Alfred North Whitehead," *Process Studies* 10, nos. 1 and 2 (Spring-Summer 1980), 44–56.

343 **"Cannot step in the same river":** Heraclitus, *Fragments*, trans. Brooks Haxton (New York: Penguin, 2003), 96.

343 **"Creative advance into novelty":** Sherburne, Donald W., *A Key to Whitehead's Process and Reality* (Chicago: University of Chicago Press, 1981), 33.

343 **"Fallacy of misplaced concreteness":** Whitehead, Alfred North, *Science and the Modern World* (New York: Free Press, 1967), 51.

352 **"The whole universe is an advancing":** Whitehead, Alfred North, *Adventures of Ideas* (New York: Free Press, 1967), 197.

352 **"The many become one":** Whitehead, Alfred North, *Process and Reality* (New York: Simon & Schuster, 1979), 21.

352 **"whole antecedent world conspires":** Whitehead, *Adventures of Ideas*, 198.

352 **"The antecedent environment":** Whitehead, Alfred North, *Modes of Thought* (New York: Simon & Schuster, 1968), 164.

353 **"Numerical models work particularly well":** Horgan, John, "From Complexity to Perplexity," *Scientific American*, June 1995, 107.

355 **"theological reflection on the spirit":** Clayton, Philip, *Adventures in the Spirit* (Minneapolis, Minn.: Fortress Press, 2008), 142.

355 **"Clayton astonished me":** Dennett, Daniel, in a report on the 2009 Darwin Celebration at Cambridge University, published by Jerry

Coyne on his blog "Why Evolution Is True," accessed October 2011, www.whyevolutionistrue.wordpress.com/2009/07/09/almost-live-report-daniel-dennett-at-the-cambridge-science-and-faith-bash/.

356   **"We see in the natural world":** Clayton, *Adventures in the Spirit*, 87.

358   **"We have a God-shaped hole":** Haught, John, "A God-Shaped Hole at the Heart of our Being," *What Is Enlightenment?* Issue 35, January–March 2007, 104.

362   **"Evolutionary theology suggests":** Haught, John F., *Making Sense of Evolution* (Louisville, Ky.: Westminster John Knox Press, 2010), 146.

362   **"Too often, we have thought":** Ibid., 147.

362   **"The world must have a god":** Teilhard de Chardin, Pierre, *Letters from a Traveller* (New York: Harper & Row, 1968), 168.

363   **"The future is the primary dwelling place":** Haught, *Making Sense of Evolution*, 138.

CHAPTER 16

365   **At the age of fourteen . . . She went on to build a life:** Houston, Jean, "Orchestrating Our Many Selves," *What Is Enlightenment?* Issue 15, Spring/Summer 1999, 108–109.

Index

Adams, Douglas, 304
Afghanistan, 229
aging, 147–49
agreements, 174–75, 178, 184
anger, in literature, 203–4
anthropocentrism, 69, 308, 314
Aquinas, Thomas, 357–58
archaic consciousness, 198, 200, 218
Archimedes, 239
art, 199, 201
    perspective in, 199, 203
Arthur, W. Brian, 144
Artilect War, The (de Garis), 150
Arya, 328
atheism, 223, 267, 268, 339, 355
Augustine, Saint, 358
Aurobindo, Sri, 148–49, 208, 236–37, 245,
    272–73, 275–78, 280, 282, 300, 322,
    326–32, 336, 349–50, 362, 371
    beginning of religious career, 326–27
    Cohen and, 331–32, 333, 334, 336
    "gnostic being" idea of, 330
    The Life Divine, 40, 233, 276, 328
    political activities of, 326, 327, 328
    The Synthesis of Yoga, 328
    Teilhard compared with, 328–29
    upbringing of, 326
Axial Age, 140, 176, 204, 311
Aztec civilization, 206–7

bacteria, 13, 50–54, 56–59, 64, 65, 69, 70,
    73, 82
Baldwin, James Mark, 27, 192–95, 198,
    214, 245, 371
Baldwin effect, 192–93
Barlow, Connie, 291
Barnett, Thomas, 96
basketball, 49–50
Beck, Don, 211–13, 215–21, 224–28, 230,
    267
beetles, 92
Behe, Michael, 120–22
Ben-Jacob, Eshel, 54
Bergson, Henri, 12, 28–29, 42, 74, 122,
    157–58, 275, 282, 306, 335, 342, 371
Berry, Thomas, 307–11
Better Angels of Our Nature, The (Pinker),
    97
Between Reason and History (Owen), 181
Beyond God (Kardong), 265
Big Bang, 253, 335
Biocosm (Gardner), 115
biology, 26–27, 190, 194, 353
    technology and, 144–45
Bloom, Howard, 57, 63, 105–12, 114, 117
body, scientific exploration of, 171, 172,
    173
Borg, Björn, 45
brain, 128, 141, 147, 211

brain (*cont.*)
  consciousness and, 159, 160, 165, 166,
    169, 172, 182, 183, 242, 288
  plasticity of, 27–28, 211, 254–55
  size of, 241–42
Brand, Stewart, 102
Brief History of Everything, A (Wilber), 234
Brooks, David, 16–17
Bruteau, Beatrice, 99
Buckle, Henry Thomas, 365
Buddhism, 159, 208, 275, 279–80, 319
Bush, George W., 21–22

California Institute of Integral Studies,
    305, 331
Cambrian period, 139
capacities, new, 302, 303
carbon, 63
categorical imperative, 259
Catholicism, 223, 267, 269–71, 273–75,
    328, 339
cellular automata, 116–17
Center for Consciousness Studies, 71–72
Center for Process Studies, 341
Center Leo Apostel, 22, 81*n*
Chalice and the Blade, The (Eisler), 98–99
Chalmers, David, 145–46
Chang Tzu, 200
Chaudri, Haridus, 331
chimpanzees, 60–61, 353
China, 197
Civil War, 225
Claremont School of Theology, 340–41
Clayton, Philip, 341, 345–51, 354–56,
    359–62
Cohen, Andrew, 158, 318–22, 331–36, 362
  Aurobindo and, 331–32, 333, 334, 336
  Teilhard and, 333, 334
  Wilber and, 333–34
Collected Works of Ken Wilber, The
    (Wilber), 238
"Coming Technological Singularity,
    The" (Vinge), 132
competition and selfishness, 62–63, 82,
    86
complex adaptive systems, 112

complexity, 37–38, 104, 113–21, 172,
    187–88, 237, 349, 353, 356
  consciousness and, 165–66, 172
compound interest, 133–34
Congo, 224
conscious evolution, 285–315, 336
Conscious Evolution (Hubbard), 289
consciousness, 35, 55, 71, 101, 155–79,
    243, 244
  archaic, 198, 200, 218
  birth of, 166–68, 183–84, 314
  brain and, 159, 160, 165, 166, 169, 172,
    182, 183, 242, 288
  collective, 160–61, 163, 165–66, 168,
    174, 184–92, 195–98
  complexity and, 165–66, 172
  Flintstones fallacy and, 181–82, 199, 266
  integral, 198–99, 207, 208, 217, 220
  magic, 198, 200–203, 205, 206, 208,
    218–19, 225–26
  matter and, 157–59
  mental-rational, 198, 204–5, 206
  mythic, 198, 202–5, 206, 208, 267
  plasticity of, 195
  power over matter through, 148
  pure, 159
consciousness, evolution of, 160–61, 163,
    181–209
  psychological development, 27, 185–
    87, 192–97
  stages of, 184–91, 195–97, 212
  see also cultural evolution
Constant Battles: Why We Fight (LeBlanc),
    97
Conway Morris, Simon, 74, 78
Cook-Greuter, Susanne, 245
cooperation, 49–70, 81–83, 86, 89–90
  governance and, 83–85
Corning, Peter A., 49
corporations, 170
cosmology, new, 302–3
Cosmos, xii, 104
countercultural movement, 176–77
Cowan, Chris, 216, 218
Coyne, Jerry, 91
creatheism, 294

creation myths, 19, 102
Creative Evolution (Bergson), 12, 275
creativity and novelty, 101–24, 349–54, 357
crises, new, 302–3
Crucible of Consciousness, The (Torey), 288
cultural evolution, 15–16, 87–88, 92–99, 139, 161, 164, 168, 170–71, 174, 176, 178, 179, 184–92, 195–98
   moral, 94–96, 259
   Spiral Dynamics and, *see* Spiral Dynamics
   see also consciousness, evolution of
culture, 160–61, 170, 197
   brain evolution and, 242
   corporate, 170
   worldviews and, *see* worldviews
culture wars, 223
Cuomo, Mario, 269–71

Dark Ages, 92–93, 95
Darwin, Charles, 8–9, 61, 73–74, 172, 330, 368
   dancing ritual witnessed by, 263–64
   On the Origin of Species, 3, 5, 17, 21, 42, 275
Darwin's Black Box (Behe), 120–21
Davies, Paul, 253
Dawkins, Richard, 62, 220, 222–23
Deacon, Terrence, 101
de Garis, Hugo, 131, 149–50
de Grey, Aubrey, 147
Delbanco, Andrew, 282
Dembski, William, 121
democracy, 229–30
Dennett, Daniel, 347, 355
depth perception, 42
Descartes, René, 34–36, 159
determinism, 110–11, 349–50, 352
Dewey, John, 74
   The Influence of Darwin on Philosophy, 21
directionality, 71–100
diversity, 20, 55, 114n, 117
Divine Milieu, The (Teilhard de Chardin), 270

DNA, 115, 123, 139, 187
Dobzhansky, Theodore, 3, 5, 10–11, 73, 190
Donald, Merlin, 211
Dowd, Michael, 93, 289–96
Durant, Will, 8–9
dust devils, 28–29

Earth, picture from space, 248
earthquakes, 26
economic structures, 243, 244
Eddington, Arthur, 119
Einstein, Albert, 133, 239–40, 360
Eisendrath, Craig, 29, 33, 34
Eisler, Riane, 98–99
embryonic development, 123
emergence, 355–57
Emerson, Ralph Waldo, 246, 263
empathy, 311
energy crisis, 137
Enlightenment, European, 8, 35, 162, 164, 176, 178, 185, 188, 189, 204, 258, 370
enlightenment, evolutionary, 317–37
EnlightenNext (formerly *What Is Enlightenment?*), xiii, xv, 123, 213, 258, 278, 318, 321, 332–34
Enríquez, Juan, 330
Esalen Institute, 331
eugenics, 14, 61
eukaryotes, 50, 56–59
Ever-Present Origin, The (Gebser), 199, 208
evolution, 7–20
   conscious, 285–315, 336
   of consciousness, *see* consciousness, evolution of
   as context for examining human nature and culture, 13–14, 18
   of culture, *see* cultural evolution
   cycle of, 55–56
   debate over, xi–xii, xiv–xv, 8
   edges and, 56
   evolution of, 139
   geological, 184
   importance of discovery of, 8–9
   as metaconcept, 368

evolution (*cont.*)
    of morality, 94–96, 259
    natural selection in, 8, 12–13, 91, 115, 116, 120, 156–57, 193, 313
    random mutation in, 8, 12–13
    reality of, 7–8
    spirituality and, *see* spirituality
    stress as creator of, 59, 93, 112–13
    of worldviews, 175–76
Evolution, 75
Evolution, Complexity, and Cognition Group (ECCO), 81
Evolutionaries, 10, 30, 31–46, 247, 283, 318, 321, 336, 341, 367, 368, 371
    as generalists, 32–41
    optimism of, 32, 43–46, 354
    time as viewed by, 32, 41–43
evolutionary enlightenment, 317–37
Evolutionary Enlightenment: A New Path to Spiritual Awakening (Cohen), 334
evolutionary impulse, 318, 332, 335
"Evolutionary Manifesto, The: Our Role in the Future Evolution of Life" (Stewart), 85
evolutionary psychology, 17–18
evolutionary worldview, new, 6, 13, 15, 16, 20, 24–30, 237–38, 336, 369–71
Evolution for Everyone (Wilson), 7, 76–77
evolutionists, 31–32
Evolution of God, The (Wright), 78, 91
Evolution of the Notion of God in Modern Thought, The (Clayton), 348
Evolution's Arrow: The Direction of Evolution and the Future of Humanity (Stewart), 72
Evolution's Purpose (McIntosh), 192
Evolving Self, The (Kegan), 195
exponential growth, 133–39
extinctions, 93, 312
eyesight, 312

Faith Instinct, The (Wade), 264
fascism, 14–15, 74, 191
fetal development, 196
Fichte, Johann, 188–89, 191
fishermen, 81, 82

flex flow, 220
Flintstones fallacy, 181–82, 199, 266
food, 137–38
football, 213–14
Foucault, Michel, 249
Foundation for Conscious Evolution, 303
Foundation for the Future, 303
four quadrants, 240–46
Fowler, James W., 267
Free University of Brussels, 22, 81
Friedman, Thomas, 89
Freud, Sigmund, 16
Freudian psychology, 18
Fuller, Buckminster, 297, 300, 330
fundamentalism, 246
Future of Man, The (Teilhard de Chardin), 25
Future of the Body, The (Murphy), 149
futurism, 297, 303
Futurist, 215

Galileo, 110
game theory, 88
Gandhi, Mohandas K., 222, 322, 326
Garden of Eden myth, 96–98
Gardner, James N., 39, 115, 117, 123, 124, 253
Gebser, Jean, 39–40, 186, 198–209, 212, 214, 218, 228, 236–37, 245, 246, 267, 371
Gell-Mann, Murray, 37
generalists, 32–41
Genial Gene, The: Deconstructing Darwinian Selfishness (Roughgarden), 62
Genius of the Beast, The (Bloom), 105
genome, 192–93
    Human Genome Project, 138
geological evolution, 184
German idealists, 186, 188–91, 330
Gibson, William, 126–27, 130, 133, 141
Gilligan, Carol, 194, 245
Global Brain, The (Bloom), 105
glucose gradients, 56, 57
gnostic being, 330
God, 362–63
    evolution of, 339–63

nature and, 360, 362, 363
panentheism and, 359–60
perfection of, 358–59
see also religion; spirituality
God Problem, The (Bloom), 108, 109
Godwin, Robert, 19, 166, 167, 222
Goethe, Johann Wolfgang von, 93, 188
Goodall, Jane, 60–61, 63
Gould, Stephen Jay, 77
governance, 83–85
Grassie, William, 355
Graves, Clare, 212–13, 215–22, 224–26, 246, 267
Greece, ancient, 164, 204
ground, 360
Gurdjieff, George, 79, 285

Habermas, Jürgen, 174, 194, 196–97, 228, 245
habits, 252–59
brain plasticity and, 254–55
moral imperative and, 257–59
morphic fields, 254–55, 257
Haeckel, Ernst, 196
Haldane, J. S., 75, 92
Halverson, William H., 24
Hartshorne, Charles, 341, 343–44
Haught, John, 122, 146, 358, 361–63
Making Sense of Evolution, 71
Heehs, Peter, 327
Hegel, Georg Wilhelm Friedrich, 9, 14, 37, 181, 186, 188–91, 193, 212, 223, 228, 342, 348–49
Heidegger, Martin, 101
Heraclitus, 343
heroic system, 217, 219
Hidden Heart of the Cosmos, The (Swimme), 167–68
"Hierarchic Logic of Emergence, The" (Deacon), 101
hierarchies, 65–69, 84, 85, 238–40
four quadrants and, 240–46
Hinduism, 102, 148, 271–72, 273, 275, 328–29
Vedanta, 319, 323
history, 214, 230

recorded, 202–3
History of Civilization in England (Buckle), 365
Hitchhiker's Guide to the Galaxy, The (Adams), 304
Hitler, Adolf, 14, 74, 76
Horgan, John, 353
horizontal gene transfer, 13
Houston, Jean, 160, 365–66
Hubbard, Barbara Marx, 289, 297–304, 330
Hubbard, Earl, 298–99
Human Genome Project, 138
"Human Nature Prepares for Momentous Leap" (Graves), 215
Huxley, Julian, 3–6, 73, 126, 371
Transhumanism, 31, 125, 126
Huxley, Thomas Henry, 3–4
"Hymn to Matter" (Teilhard de Chardin), 274

idealism, 230, 278, 351, 354
Iliad (Homer), 183, 204
immortality, 148–49
India, 197
Influence of Darwin on Philosophy, The (Dewey), 21
information, 144–47
In Over Our Heads (Kegan), 195
integral consciousness, 198–99, 207, 208, 217, 220
Integral Consciousness and the Future of Evolution (McIntosh), 162
Integral Institute, 176
integral philosophy, 233–59
four quadrants in, 240–46
Integral Spirituality (Wilber), 170
integral yoga, 237, 329
integration, 22, 37–41, 114n
intelligence, 68, 73, 77, 94, 109, 141–44, 304
artificial, 128–30, 132, 150
intelligent design, 121
Internet, 141
intersubjectivity, 164, 168–72, 178, 249
Investigations (Kauffman), 313–14

Iraq, 229
Iron Age, 176
Islamic fundamentalism, 208

James, William, 251, 282, 341
Jaynes, Julian, 155, 183–84
Jefferson, Thomas, 162, 164
John XXIII, Pope, 270
Johnson, Steven, xiv
Joyce, James, 40
Judaism, 271
Jung, Carl, 168, 173, 323

Kant, Immanuel, 181, 259
Kardong, Kenneth V., 265
Kashmir Shaivism, 271–72
Kauffman, Stuart, 13, 38, 115–16, 119,
    313–14, 360
Kegan, Robert, 188, 195, 196
Kelly, Kevin, 113, 128, 140
knowledge, 311, 312
Kohlberg, Lawrence, 194, 196, 216, 245,
    267
Kosmos, 240, 244
Krugman, Paul, 34
Kurzweil, Ray, 132–40, 142–43, 145–48,
    151

Lachman, Gary, 197–98
lactose intolerance, 193
language, 175, 201, 203
Laplace, Pierre-Simon, 110–11, 349
LeBlanc, Steven, 96–97
Letter to Ouspensky (Gurdjieff), 285
life, 313–14
Life Divine, The (Aurobindo), 40, 233,
    276, 328
life expectancy, 147–49
Life's Solution: Inevitable Humans in a
    Lonely Universe (Conway Morris),
    78
Live Long Enough to Live Forever
    (Kurzweil), 147
Loevinger, Jane, 194, 216, 245
Lucifer Principle, The (Bloom), 105
Lyotard, Jean-François, 234

magic consciousness, 198, 200–203, 205,
    206, 208, 217–19, 225–26
Maharshi, Ramana, 320, 322–26, 331
Making Sense of Evolution (Haught), 71
Margulis, Lynn, 13, 50–52, 57–58, 59, 62,
    65, 67, 68, 82
marriage, 17
Marx, Karl, 8–9, 14, 16, 243, 266, 298
Marxism, 14, 74, 244
Maslow, Abraham, 213, 299
Mathematica Principia (Whitehead and
    Russell), 341
matter, 205
    consciousness and, 157–59
Maugham, W. Somerset, 323
Mayr, Ernst, 3, 5, 74–75
McCracken, Grant, 170
McEnroe, John, 45
McIntosh, Steve, 162–65, 169–71, 174,
    176–78, 189, 192, 196, 236
McLuhan, Marshall, 245
Mead, Margaret, 299
meaning, 15
medicine, 138
meditation, 156, 158–60, 179, 280
Meeker, Tobias, 293
Melville, Herman, 301
Men in Black, 64
mental-rational consciousness, 198,
    204–5, 206
Merton, Thomas, 300
Metanexus, 355
Michelangelo, 171
Mid-West Quarterly, 252
mind, 182
    see also consciousness
Mind So Rare, A (Donald), 211
modernism, 24, 162, 197, 204, 217, 219,
    222–24, 229–30, 231, 267
monkeys/typewriter metaphor, 108–9
Montezuma, 206
Moore, Gordon, 137
Moore's Law, 136–37
Moral Animal, The (Wright), 78
morality:
    categorical imperative and, 259

evolution of, 94–96, 259
grooves and, 257–59
morphic fields, 254–55, 257
movement, 25–30, 182–83, 342–43, 367, 368
Muir, John, 67–68
Murphy, Michael, 92, 149, 331
mythic consciousness, 198, 202–5, 206, 208, 267

Nash, Steve, 49–50
National Center for Science Education, 120
Native Americans, 226
naturalism, 267, 268, 363
nature, 67–68, 204, 207, 310
  God and, 360, 362, 363
  laws of, 252–53
Nature of Technology, The: What It Is and How It Evolves (Arthur), 144
Neuromancer (Gibson), 126–27, 141
neuroscience, 27–28
New Orleans, La., 56, 226
New Republic, 91
New Science of Life, A (Sheldrake), 254, 255
Newton, Isaac, 110, 171, 239
New York Times, 16, 34, 264, 340
New York Times Magazine, 90
Niebuhr, Reinhold, 301
9/11 attacks, 21–22, 69, 229
No Free Lunch: Why Specified Complexity Cannot Be Purchased Without Intelligence (Dembski), 121
No Maps for These Territories, 133
Nonzero: The Logic of Human Destiny (Wright), 78, 87, 95
non-zero-sum interactions, 88–90, 95
noosphere, 164–66, 168, 169, 172, 174, 178–79, 255
Norse mythology, 19
novelty and creativity, 101–24, 349–54, 357

Obitts, Stanley, 346
objectivity, 250
Omnipotence and Other Theological Mistakes (Hartshorne), 344
"On the Origin of Mitosing Cells" (Margulis), 51–52
On the Origin of Species (Darwin), 3, 5, 17, 21, 42, 275
optimism, 32, 43–46, 76, 130, 148, 354
Origin of Consciousness in the Breakdown of the Bicameral Mind, The (Jaynes), 155, 183–84
origin stories, 19, 102
Out of Darkness (Torey), 285–86
Owen, David, 181, 185
oxygen, 57, 58

panentheism, 359–60
Pannenberg, Wolfhart, 347–48, 351
pantheism, 294, 360, 363
Parliament of the World's Religions, 87–88
Pascal, Blaise, 118
Patten, Terry, 237
Paul, Saint, 301
Peirce, Charles Sanders, 251–53, 351
perspectives, 247–51
Phenomenology of Spirit, The (Hegel), 189
Phenomenon of Man, The (Teilhard de Chardin), 4, 167, 275, 299–300, 307, 366
Philosophical Inquiries into the Nature of Human Freedom (Schelling), 339
photosynthesis, 57
physics, 353
  laws of, 252–53
  quantum, 34, 113, 349, 351
Piaget, Jean, 192, 194–96, 245
Picasso, Pablo, 199
Pinker, Steven, 97
Plato, 279, 359
Plotkin, Henry, 193
Plutarch, xv
Poonja, H.W.L., 320
Pope, Alexander, 61
Popper, Karl, 79, 191
postmodernism, 76, 217, 219, 221–24, 231, 234, 267
Power of Now, The (Tolle), 278–79

predictability, 110–11, 116, 130, 349–50
process philosophy, 342
Progress: Can We Do Without It?
(Salvadori), 76
progress and directionality, 71–100
psychology, 173
developmental, 27, 185–87, 192–97
evolutionary, 17–18
Freudian, 18
Spiral Dynamics and, *see* Spiral
Dynamics

quadrants, four, 240–46
quantum physics, 34, 113, 349, 351

Ray, Paul, 245
Razor's Edge, The (Maugham), 323
Real American Dream, The (Delbanco), 282
"Recent Changes in the Structure of
Consciousness" (Sleutels), 181–82
reductionism, 36, 38, 60–61, 67, 77, 116,
244
Reinventing the Sacred (Kauffman), 13, 116
relationships, 169–70
religion, 223–24, 264–68, 281–82, 337
antiworld bias in, 270–71, 277, 279–81
Axial Age and, 140, 176, 204, 311
Buddhism, 159, 208, 275, 279–80, 319
Cartesian revolution and, 36
Catholicism, 223, 267, 269–71, 273–75,
328, 339
conversion and, 242–43, 291–92
diversity of practice in, 266–68
evolutionary theology, 339–63
evolution debate and, xi–xii, xiv–xv, 8
Hinduism, *see* Hinduism
see also spirituality
Renaissance, 258
artistic perspective and, 199
and scientific exploration of body, 171,
172, 173
revelations, 295
Richard, Mirra, 330
Richards, Robert, 189, 198
Rolston, Holmes, III, 313, 314
Roman Empire, 56, 93, 95

Romantics, 67, 273, 310, 360
Roughgarden, Joan, 62
Rousseau, Jean-Jacques, 97
Ruse, Michael, 62
Russell, Bertrand, 251, 298, 341–42
Rwanda, 224

Sagan, Carl, xii, 50, 104–5, 306, 360
Sahtouris, Elisabet, 52–55, 58–59, 62,
65, 82
Salamon, Anna, 125
Salk, Jonas, 300
Salvadori, Massimo, 76
Santa Fe Institute, 37
Schelling, Friedrich, 188–89, 191, 348–49,
360
*Philosophical Inquiries into the Nature of
Human Freedom*, 339
Schopenhauer, Arthur, 191
science, 250, 268, 367–68
Scott, Eugenie, 120
second law of thermodynamics, 118–20
Secret, The, 278
self, 35
selfishness and competition, 62–63, 82,
86
Sex, Ecology, Spirituality (Wilber), 238
Shankara, 319
Sheldrake, Rupert, 253–55, 351
Shelley, Percy Bysshe, 310
Sherif, Muzafer, 225
singularity, 129–32
Singularity Is Near, The (Kurzweil), 132,
136
Singularity Summit, 125–26, 127–28, 130
Skinner, B. F., 105
Skunk Works, 56
Sleutels, Jan, 181–83
Smith, Huston, 279
social Darwinism, 14, 61, 187
solar energy, 137
solidity, 9, 21–30, 46, 174, 182–83, 189,
195, 252, 253, 266, 342–43, 363,
368, 369
Sorokin, Pitirim, 236
South Africa, 225–28

INDEX

"Sovereignty of Ethics, The" (Emerson), 263

specialization, 33–38

Spectrum of Consciousness, The (Wilber), 234

Spinoza, Baruch, 360

Spiral Dynamics, 194, 211–31, 245
  healthy and unhealthy expressions of systems in, 221, 230
  levels of, 217–22
  South African culture and, 225–28

spirituality, 16, 139–40, 172–73, 305, 325, 336–37
  evolutionary, 263–83, 355, 363
  evolutionary enlightenment, 317–37
  transcendence and, 272, 278, 282
  see also religion

Stages of Faith (Fowler), 267

steady-state cosmos, 27

Steiner, Rudolf, 74

Stewart, John, 72, 73, 75, 77–86, 91, 95, 99

stress, 59, 93, 112–13

survivalist system, 217, 218, 225

Swimme, Brian, 20, 167–68, 289, 290, 305–6, 360
  Cohen and, 333, 334

symbiogenesis, 50–51

Synergism Hypothesis, The (Corning), 49

Synthesis of Yoga, The (Aurobindo), 328

systems science, 37–38

Tao Te Ching, 19

Tarnas, Richard, 191

Taylor, Charles, 35

technology, 15, 303
  biology and, 144–45
  singularity and, 129–32
  transhumanism and, 125–51

Teilhard de Chardin, Pierre, 11, 42, 79, 82, 84, 88, 118, 141, 164–66, 168, 208, 272–75, 282, 299–300, 307–9, 330, 344, 349–50, 361–62, 370, 371
  Aurobindo compared with, 328–29
  Cohen and, 333, 334
  The Divine Milieu, 270
  The Future of Man, 25

Houston and, 365–66
  "Hymn to Matter," 274
  life of, 273–74
  The Phenomenon of Man, 4, 167, 275, 299–300, 307, 366

teleology (directionality), 71–100

theology, 339–40
  evolutionary, 339–63
  see also religion

Theory of Everything, A (Wilber), 234

thermodynamics, second law of, 118–20

Thompson, William Irwin, 65, 198

Thoreau, Henry David, 67–68

Tillich, Paul, 159, 258, 360

time, 203
  evolutionary, 32, 41–43

Tolle, Eckhart, 278–79

Torey, Zoltan, 285–89

"Toward a Science of Consciousness" conference, 71–72, 77, 145–46, 181

traditionalism, 217, 219, 222–24, 231, 267

transcendence, 272, 278, 282

transhumanism, 125–51, 309

"Transhumanism" (Huxley), 31, 125, 126

truth, 37, 189–90

Turner, Fred, 189n

Tyson, Neil deGrasse, 306–7

universe, 311
  deterministic view of, 110–11, 349–50, 352

Universe Is a Green Dragon, The (Swimme), 310

Universe Story, The (Swimme and Berry), 310

values, 215
  Spiral Dynamics and, *see* Spiral Dynamics

Vedanta, 319, 323

vegetarianism, 65–67

Vernadsky, Vladimir, 165

Vinge, Vernor, 130, 132

Voltaire, 31, 188, 223

von Neumann, John, 88, 131–32

Wade, Nicholas, 264

Walter, Chip, 330

war and violence, 96–99
  world wars, 74, 75–76, 79, 94, 95

water, 356

Watts, Alan, 66

Wharton, Edith, 301

What Is Enlightenment?, see EnlightenNext

White, John, 330

Whitehead, Alfred North, 29, 251, 252,
        282, 325, 341–44, 351–52, 354, 371

Wilber, Ken, 22, 38, 62, 169, 170, 176–77,
        190, 220–21, 228, 234–48, 251,
        254–59, 267, 317, 331, 351
  Cohen and, 333–34

Wilson, David Sloan, 7, 76–77

Winfrey, Oprah, 278

Wired, 113

Wolfram, Stephen, 116–17

Wordsworth, William, 97–98, 310

World Future Society, 303

worldviews, 16, 21–24, 69–70, 161, 163,
        174–78, 184–87, 189–90, 195–96,
        212, 249
  agreements in, 174–75, 178, 184
  evolution of, 175–76
  evolutionary, new, 6, 13, 15, 16, 20,
        24–30, 237–38, 336, 369–71
  in Spiral Dynamics, *see* Spiral
        Dynamics
  touchstone propositions of, 24–25

World War I, 74, 75–76, 79, 94, 95, 204,
        273–74

World War II, 74, 75–76, 79, 94, 95, 204,
        229, 297–98

Wright, N. T., 23

Wright, Robert, 78, 87–91, 93–94, 95, 99,
        115, 137, 156–57

zero-sum interactions, 88

Zulus, 225–26